T0320420

The Science of Lithium

The use of lithium is increasing at breathtaking speed and is currently changing key industries and the way people live. Lithium is used in an ever-growing number of electric vehicles (EVs), in laptops where the battery power lasts all day on a single charge, in solar panels mounted on roofs, and several other uses, all of which are discussed in this book. *The Science of Lithium* fills a wide gap of information previously missing from other published texts dealing with the green energy revolution currently in progress; it provides a comprehensive survey of information on this highly sought-after element, including its chemistry, metallurgical properties, and industrial applications, as well as its drawbacks and environmental implications.

The Science of Lithium

Frank R. Spellman

CRC Press
Taylor & Francis Group
Boca Raton London New York

CRC Press is an imprint of the
Taylor & Francis Group, an **informa** business

Designed cover image: Shutterstock

First edition published 2024
by CRC Press
6000 Broken Sound Parkway NW, Suite 300, Boca Raton, FL 33487-2742

and by CRC Press
4 Park Square, Milton Park, Abingdon, Oxon, OX14 4RN

CRC Press is an imprint of Taylor & Francis Group, LLC

Library of Congress Cataloging-in-Publication Data
Names: Spellman, Frank R., author.
Title: The science of lithium / Frank R. Spellman.
Description: First edition. | Boca Raton : CRC Press, 2024. |
Includes bibliographical references and index.
Identifiers: LCCN 2023004858 (print) | LCCN 2023004859 (ebook) |
ISBN 9781032482040 (hbk) | ISBN 9781032482057 (pbk) | ISBN 9781003387879 (ebk)
Subjects: LCSH: Lithium.
Classification: LCC TK2945.L58 S64 2024 (print) | LCC TK2945.L58 (ebook) |
DDC 661/.0381—dc23/eng/20230216
LC record available at https://lccn.loc.gov/2023004858
LC ebook record available at https://lccn.loc.gov/2023004859

ISBN: 978-1-032-48204-0 (hbk)
ISBN: 978-1-032-48205-7 (pbk)
ISBN: 978-1-003-38787-9 (ebk)

DOI: 10.1201/9781003387879

Typeset in Times
by codeMantra

Contents

PART 1   Introduction

PART 2   Applications

PART 3   Medicinal Use of Lithium

PART 4   Lithium Power Applications

PART 5   Environmental and Safety Concerns

PART 6 Fuel Cells

PART 7 Final Word on Environmental and Safety Concerns

Preface

The Science of Lithium is the ninth volume in the acclaimed series that includes *The Science of Electric Vehicles (EVs): Concepts and Applications, The Science of Rare Earth Elements: Concepts and Applications, The Science of Water, The Science of Air, The Science of Environmental Pollution, The Science of Renewable Energy, The Science of Waste, and The Science of Wind Power*, all of which bring this highly successful series fully into the 21st century. *The Science of Lithium* continues the series mantra based on good science and not feel-good science. It also continues to be presented in the Author's trademark conversational style—making sure communication is certain.

This practical, direct book presents technologies and techniques, as well as real-world usage of and operation of today's lithium-based applications. This book is designed to be used as an information source for the general reader, or for a course in chemistry or in renewable energy engineering fields where lithium is a key player. In this book, you will find lithium used for medicinal purposes and also basic electrical principles, physics, chemistry, the use of lithium-ion batteries for electric vehicles (EVs), and several other current applications—again, all presented in understandable down to Earth prose; no convoluted verbiage is present or allowed.

Concern for the environment and the impact of environmental pollution has brought about the trend (and the need) to shift from the use and reliance on hydrocarbons to energy-power sources that are pollution neutral or near pollution neutral and renewable. We are beginning to realize that we are responsible for much of the environmental degradation of the past and present—all of which is readily apparent today—human footprints grow in size with each passing day. Moreover, the impact of 200 years of industrialization and surging population growth has far exceeded the future supply of hydrocarbon power sources. So, the implementation of renewable energy sources is surging, and along with it there is a corresponding surge in utilization of lithium-ion batteries as a primary means of transferring from fossil fuels to renewable energy producers.

Why a text on the science of lithium? Simply put, studying physics, electricity, motion, materials, metals, products, and the unfolding global energy revolution without including the inherent science connection is analogous to attempting to reach an unknown, unfamiliar location without being able to read a map, written directions, or digital device—and today that digital device probably is powered by a lithium-ion battery.

Throughout this text, common-sense approaches and practical examples have been presented. Again, because this is a science text, I have adhered to scientific principles, models, and observations. But you need not be a scientist to understand the principles and concepts presented. What is needed is an open mind, a love for the challenge of wading through all the information, an ability to decipher problems, and the patience to answer the questions relevant to each topic presented. The text follows a pattern that is nontraditional; that is, the paradigm used here is based on real-world experience, not on theoretical gobbledygook. Real-life situations are woven throughout the fabric of this text and presented in straightforward, plain English to give the facts, knowledge, and information to enable understanding and needed to make informed decisions.

Environmental issues are attracting ever-increasing attention at all levels. The problems associated with these issues are compounded and made more difficult by the sheer number of factors involved in managing any phase of any problem. Because the issues affect so many areas of society, the dilemma makes us hunt for strategies that solve the problems for all, while maintaining a safe environment without excessive regulation and cost—Gordian knots that defy easy solutions.

The preceding statement goes to the heart of why this text is needed. Presently, only a limited number of individuals have sufficient background in the science of lithium and its concepts and applications in the world of industrial and practical functions, purposes, and uses to make informed decisions on 21st century product production, usage, and associated environmental issues.

This book is designed to allow the readers to stay ahead of the curve when it comes to having knowledge of one of the major ingredients in the breathtaking revolution intended to completely transform key industries and the way we live.

The bottom line: Critical to solving real-world environmental problems is for all of us to remember that old saying, we should take nothing but pictures, leave nothing but footprints, kill nothing but time and sustain ourselves with the flow of clean, safe, renewable energy—and truck-on under lithium powered applications.

Frank R. Spellman
Norfolk, VA

Conversion Factors and SI Units

Note: *Parts per million (ppm)*—the number of parts by weight of a substance per million parts of water. This unit is commonly used to represent pollutant concentrations. Large concentrations are expressed in percentages.

Parts per billion (ppb)

Note: For comparative purposes, we like to say that 1-ppm is analogous to a full shot glass of water sitting in the bottom of a full standard swimming pool. Consider 1-ppb as one drop from a dropper into the same swimming pool.

The units most commonly used by environmental engineering professionals are based on the complicated English System of Weights and Measures. However, bench work is usually based on the metric system or the International System of Units (SI) due to the convenient relationship between milliliters (mL), cubic centimeters (cm^3), and grams (g).

The SI is a modernized version of the metric system established by International agreement. The metric system of measurement was developed during the French Revolution and was first promoted in the United States in 1866. In 1902, proposed congressional legislation requiring the U.S. Government to use the metric system exclusively was defeated by a single vote. Although we use both systems in this text, SI provides a logical and interconnected framework for all measurements in engineering, science, industry, and commerce. The metric system is much simpler to use than the existing English system since all its units of measurement are divisible by 10.

Before listing the various conversion factors commonly used in environmental engineering it is important to describe the prefixes commonly used in the SI system. These prefixes are based on the power ten. For example, a "kilo" means 1,000 g, and a "centimeter" means 1/100th of 1 m. The 20 SI prefixes used to form decimal multiples and submultiples of SI units are presented.

Note that the kilogram is the only SI unit with a prefix as part of its name and symbol. Because multiple prefixes may not be used, in the case of the kilogram, the prefix names are used with the unit's name "gram" and the prefix symbols are used with the unit symbol "g." With this exception, any SI prefix may be used with any SI unit, including the degree Celsius and its symbol °C.

SI Prefixes

Factor	Name	Symbol
10^{24}	Yotta	Y
10^{21}	Zetta	Z
10^{18}	Exa	E
10^{15}	Peta	P
10^{12}	Tera	T
10^9	Giga	G
10^6	Mega	M
10^3	Kilo	k
10^2	Hecto	h
10^1	Deka	da

10^{-1}	Deci	d
10^{-2}	Centi	c
10^{-3}	Milli	m
10^{-6}	Micro	μ
10^{-9}	Nano	n
10^{-12}	Pico	p
10^{-15}	Femto	f
10^{-18}	Atto	a
10^{-21}	Zepto	z
10^{-24}	Yocto	y

EXAMPLE 1

10−6 kg = 1 mg (one milligram), but not 10−6 kg = 1 μkg (one microkilogram)

EXAMPLE 2

Consider the height of the Washington Monument. We may write h_w = 169,000 mm = 16,900 cm = 169 m = 0.169 km using the millimeter (SI prefix "milli," symbol "m"); centimeter (SI prefix "centi," symbol "c"); or kilometer (SI prefix "kilo," symbol "k").

EXAMPLE 3

Problem: Find degrees in Celsius of water at 72°F.
   Solution:

$$°C = (F - 32) \times 5/9 = (72 - 32) \times 5/9 = 22.2$$

DID YOU KNOW?

The Fibonacci sequence is the following sequence of numbers:

$$1, 1, 2, 3, 5, 8, 13, 21, 34, 55, 89, 144, \ldots$$

Or, alternatively,

$$0.1, 1, 2, 3, 5, 8, 13, 21, 34, 55, 89, 144, \ldots$$

Two important points, the first obvious: each term from the third onward is *the sum of the previous two*. Another point to notice is that if you divide each number in the sequence by the next number, beginning with the first, an interesting thing appears to be happening:
   1/1=1, 1/2=0.5, 2/3=0.66666…, 3/5=0.6, 5/8=0.625, 8/13=0.61538…, 13/21= 0.61904 …, (the first of these ratios appears to be converging to a number just a bit larger than 0.6).

CONVERSION FACTORS

Conversion factors are given below in alphabetical order and in unit category listing order.

ALPHABETICAL LISTING OF CONVERSION FACTORS

Factors	Metric (SI) or English Conversions
1 atm (atmosphere) =	1.013 bars
	10.133 newtons/cm^2 (newtons/square centimeter)
	33.90 ft. of H$_2$O (feet of water)
	101.325 kp (kilopascals)
	1,013.25 mg (millibars)
	13.70 psia (pounds/square inch—absolute)
	760 torr
	760 mm Hg (millimeters of mercury)
1 bar =	0.987 atm (atmospheres)
	1 × 10^6 dynes/cm^2 (dynes/square centimeter)
	33.45 ft of H$_2$O (feet of water)
	1 × 10^5 pascals [nt/m^2] (newtons/square meter)
	750.06 torr
	750.06 mm Hg (millimeters of mercury)
1 Bq (becquerel) =	1 radioactive disintegration/second
	2.7 × 10^{-11} Ci (curie)
	2.7 × 10^{-8} mCi (millicurie)
1 BTU (British Thermal Unit) =	252 cal (calories)
	1,055.06 j (joules)
	10.41 liter-atmosphere
	0.293 watt-hours
1 cal (calories) =	3.97 × 10^{-3} BTUs (British Thermal Units)
	4.18 j (joules)
	0.0413 liter-atmospheres
	1.163 × 10^{-3} watt-hours
1 cm (centimeters) =	0.0328 ft (feet)
	0.394 in (inches)
	10,000 microns (micrometers)
	100,000,000 Å =10^8 Å (Ångstroms)
1 cc (cubic centimeter) =	3.53 × 10^{-5} ft^3 (cubic feet)
	0.061 in^3 (cubic inches)
	2.64 × 10^{-4} gal (gallons)
	52.18 ℓ (liters)
	52.18 mL (milliliters)
1 ft^3 (cubic foot) =	28.317 cc (cubic centimeters)
	1,728 in^3 (cubic inches)
	0.0283 m^3 (cubic meters)

(*Continued*)

Factors	Metric (SI) or English Conversions
	7.48 gal (gallons)
	28.32 ℓ (liters)
	29.92 qts (quarts)
1 in³	16.39 cc (cubic centimeters)
	16.39 ml (milliliters)
	5.79×10^{-4} ft³ (cubic feet)
	1.64×10^{-5} m³ (cubic meters)
	4.33×10^{-3} gal (gallons)
	0.0164 ℓ (liters)
	0.55 fl oz (fluid ounces)
1 m³ (cubic meter) =	1,000,000 cc = 10^6 cc (cubic centimeters)
	33.32 ft³ (cubic feet)
	61,023 in³ (cubic inches)
	264.17 gal (gallons)
	1,000 ℓ (liters)
1 yd³ (cubic yard) =	201.97 gal (gallons)
	764.55 ℓ (liters)
1 Ci (curie) =	3.7×10^{10} radioactive disintegrations/second
	3.7×10^{10} Bq (becquerel)
	1,000 mCi (millicurie)
1 day =	24 hrs (hours)
	1,440 min (minutes)
	86,400 sec (seconds)
	0.143 weeks
	2.738×10^{-3} yrs (years)
1°C (expressed as an interval) =	1.8°F = [9/5] °F (degrees Fahrenheit)
	1.8°R (degrees Rankine)
	1.0 K (degrees Kelvin)
°C (degree Celsius) =	$[(5/9)(°F - 32°)]$
1°F (expressed as an interval) =	0.556°C = [5/9]°C (degrees Celsius)
	1.0°R (degrees Rankine)
	0.556 K (degrees Kelvin)
°F (degree Fahrenheit) =	$[(9/5)(°C) + 32°]$
1 dyne =	1×10^{-5} nt (newton)
1 ev (electron volt) =	1.602×10^{-12} ergs
	1.602×10^{-19} j (joules)
1 erg =	1 dyne-centimeters
	1×10^{-7} j (joules)
	2.78×10^{-11} watt-hours
1 fps (feet/second) =	1.097 kmph (kilometers/hour)
	0.305 mps (meters/second)
	0.01136 mph (miles/hour)
1 ft (foot) =	30.48 cm (centimeters)
	12 in (inches)
	0.3048 m (meters)

(Continued)

Factors	Metric (SI) or English Conversions
	1.65×10^{-4} nt (nautical miles)
	1.89×10^{-4} mi (statute miles)
1 gal (gallon) =	3,785 cc (cubic centimeters)
	0.134 ft^3 (cubic feet)
	231 in^3 (cubic inches)
	3.785 ℓ (liters)
1 gm (gram)	0.001 kg (kilogram)
	1,000 mg (milligrams)
	1,000,000 ng = 10^6 ng (nanograms)
	2.205×10^{-3} lbs (pounds)
1 gm/cc (grams/cubic cent.) =	62.43 lbs/ft^3 (pounds/cubic foot)
	0.0361 lbs/in^3 (pounds/cubic inch)
	8.345 lbs/gal (pounds/gallon)
1 Gy (gray) =	1 j/kg (joules/kilogram)
	100 rad
	1 Sv (sievert) [unless modified through division by an appropriate factor, such as Q and/or N]
1 hp (horsepower) =	745.7 j/sec (joules/sec)
1 hr (hour) =	0.0417 days
	60 min (minutes)
	3,600 sec (seconds)
	5.95×10^{-3} weeks
	1.14×10^{-4} yrs (years)
1 in (inch) =	2.54 cm (centimeters)
	1,000 mil
I inch of water =	1.86 mm Hg (millimeters of mercury)
	249.09 pascals
	0.0361 psi (lbs/in^2)
1 j (joule) =	9.48×10^{-4} BTUs (British Thermal Units)
	0.239 cal (calories)
	10,000,000 ergs = 1×10^7 ergs
	9.87×10^{-3} liter-atmospheres
1.0	nt-m (newton-meters)
1 kcal (kilocalories) =	3.97 BTUs (British Thermal Units)
	1,000 cal (calories)
	4,186.8 j (joules)
1 kg (kilogram) =	1,000 gms (grams)
	2.205 lbs (pounds)
1 km (kilometer) =	3,280 ft (feet)
	0.54 nt (nautical miles)
	0.6214 mi (statute miles)
1 kw (kilowatt) =	56.87 BTU/min (British Thermal Units)
	1.341 hp (horsepower)
	1,000 j/sec (kilocalories)
1 kw-hr (kilowatt-hour) =	3,412.14 BTU (British Thermal Units)

(Continued)

Factors	Metric (SI) or English Conversions
	3.6×10^6 j (joules)
	859.8 kcal (kilocalories)
1 ℓ (liter) =	1,000 cc (cubic centimeters)
	1 dm^3 (cubic decimeters)
	0.0353 ft^3 (cubic feet)
	61.02 in^3 (cubic inches)
	0.264 gal (gallons)
	1,000 ml (milliliters)
	1.057 qts (quarts)
1 m (meter) =	1×10^{10} Å (Ångstroms)
	100 cm (centimeters)
	3.28 ft (feet)
	39.37 in (inches)
	1×10^{-3} km (kilometers)
	1,000 mm (millimeters)
	1,000,000 μ = 1×10^6 μ (micrometers)
	1×10^9 nm (nanometers)
1 mps (meters/second) =	196.9 fpm (feet/minute)
	3.6 kmph (kilometers/hour)
	2.237 mph (miles/hour)
1 mph (mile/hour) =	88 fpm (feet/minute)
	1.61 kmph (kilometers/hour)
	0.447 mps (meters/second)
1 kt (nautical mile) =	6,076.1 ft (feet)
	1.852 km (kilometers)
	1.15 mi (statute miles)
	2,025.4 yds (yards)
1 mi (statute mile) =	5,280 ft (feet)
	1.609 km (kilometers)
	1,609.3 m (meters)
	0.869 nt (nautical miles)
	1,760 yds (yards)
1 miCi (millicurie) =	0.001 Ci (curie)
	3.7×10^{10} radioactive disintegrations/second
	3.7×10^{10} Bq (becquerel)
1 mm Hg (mm of mercury) =	1.316×10^{-3} atm (atmosphere)
	0.535 in H$_2$O (inches of water)
	1.33 mb (millibars)
	133.32 pascals
	1 torr
	0.0193 psia (pounds/square inch—absolute
1 min (minute) =	6.94×10^{-4} days
	0.0167 hrs (hours)
	60 sec (seconds)
	9.92×10^{-5} weeks

(Continued)

Factors	Metric (SI) or English Conversions
	1.90×10^{-6} yrs (years)
1 N (newton) =	1×10^5 dynes
1 N-m (newton-meter) =	1.00 j (joules)
	2.78×10^{-4} watt-hours
1 ppm (parts/million-volume) =	1.00 mL/m³ (milliliters/cubic meter)
1 ppm [wt] (parts/million-weight) =	1.00 mg/kg (milligrams/kilograms)
1 pascal =	9.87×10^{-6} atm (atmospheres)
0.01 mb (millibars)	4.015×10^{-3} in H_2O (inches of water)
	7.5×10^{-3} mm Hg (milliliters of mercury)
1 lb (pound) =	453.59 g (grams)
	16 oz (ounces)
1 lbs/ft³ (pounds/cubic foot) =	16.02 g/l (grams/liter)
1 lbs/ft³ (pounds/cubic inch) =	27.68 gms/cc (grams/cubic centimeter)
	1,728 lbs/ft³ (pounds/cubic feet)
1 psi (pounds/square inch) =	0.068 atm (atmospheres)
	27.67 in H_2O (inches or water)
	68.85 mb (millibars)
	51.71 mm Hg (millimeters of mercury)
	6,894.76 pascals
1 qt (quart) =	946.4 cc (cubic centimeters)
	57.75 in³ (cubic inches)
	0.946 ℓ (liters)
1 rad =	100 ergs/gm (ergs/gram)
	0.01 Gy (gray)
	1 rem [unless modified through division by an appropriate factor, such as Q and/or N]
1 rem	1 rad [unless modified through division by an appropriate factor, such as Q and/or N]
1 Sv (sievert) =	1 Gy (gray) [unless modified through division by an appropriate factor, such as Q and/or N]
1 cm² (square centimeter)	$= 1.076 \times 10^{-3}$ ft² (square feet)
	0.155 in² (square inches)
	1×10^{-4} m² (square meters)
1 ft² (square foot) =	2.296×10^{-5} acres
	9.296 cm² (square centimeters)
	144 in² (square inches)
	0.0929 m² (square meters)
1 m² (square meter) =	10.76 ft² (square feet)
	1,550 in² (square inches)
1 mi² (square mile) =	640 acres
	2.79×10^7 ft² (square feet)
	2.59×10^6 m² (square meters)
1 torr =	1.33 mb (millibars)
1 watt =	3.41 BTI/hr (British Thermal Units/hour)

(Continued)

Factors	Metric (SI) or English Conversions
	1.341×10^{-3} hp (horsepower)
	52.18 j/sec (joules/second)
1 watt-hour =	3.412 BTUs (British Thermal Unit)
	859.8 cal (calories)
	3,600 j (joules)
	35.53 liter-atmosphere
1 week =	7 days
	168 hrs (hours)
	10,080 min (minutes)
	6.048×10^5 sec (seconds)
	0.0192 yrs (years)
1 yr (year) =	365.25 days
	8,766 hrs (hours)
	5.26×10^5 min (minutes)
	3.16×10^7 sec (seconds)
	52.18 weeks

CONVERSION FACTORS BY UNIT CATEGORY

	Units of Length
1 cm (centimeter) =	0.0328 ft (feet)
	0.394 in (inches)
	10,000 microns (micrometers)
	$100,000,000$ Å $= 10^8$ Å (Ångstroms)
1 ft (foot) =	30.48 cm (centimeters)
	12 in (inches)
	0.3048 m (meters)
	1.65×10^{-4} nt (nautical miles)
	1.89×10^{-4} mi (statute miles)
1 in (inch) =	2.54 cm (centimeters)
	1,000 mil
1 km (kilometer) =	3,280.8 ft (feet)
	0.54 nt (nautical miles)
	0.6214 mi (statute miles)
1 m (meter) =	1×10^{10} Å (Ångstroms)
	100 cm (centimeters)
	3.28 ft (feet)
	39.37 in (inches)
	1×10^{-3} km (kilometers)
	1,000 mm (millimeters)
	$1,000,000$ μ $= 1 \times 10^6$ μ (micrometers)

(*Continued*)

Units of Length

	1×10^9 nm (nanometers)
1 kt (nautical mile) =	6,076.1 ft (feet)
	1.852 km (kilometers)
	1.15 km (statute miles)
	2.025.4 yds (yards)
1 mi (statute mile) =	5,280 ft (feet)
	1.609 km (kilometers)
	1.690.3 m (meters)
	0.869 nt (nautical miles)
	1,760 yds (yards)

Units of Area

1 cm^2 (square centimeter) =	1.076×10^{-3} ft^2 (square feet)
	0.155 in^2 (square inches)
	1×10^{-4} m^2 (square meters)
1 ft^2 (square foot) =	2.296×10^{-5} acres
	929.03 cm^2 (square centimeters)
	144 in^2 (square inches)
	0.0929 m^2 (square meters)
1 m^2 (square meter) =	10.76 ft^2 (square feet)
	1,550 in^2 (square inches)
1 mi^2 (square mile) =	640 acres
	2.79×10^7 ft^2 (square feet)
	2.59×10^6 m^2 (square meters)

Units of Volume

1 cc (cubic centimeter) =	3.53×10^{-5} ft^3 (cubic feet)
	0.061 in^3 (cubic inches)
	2.64×10^{-4} gal (gallons)
	0.001 ℓ (liters)
	1.00 ml (milliliters)
1 ft^3 (cubic foot) =	28,317 cc (cubic centimeters)
	1,728 in^3 (cubic inches)
	0.0283 m^3 (cubic meters)
	7.48 gal (gallons)
	28.32 ℓ (liters)
	29.92 qts (quarts)
1 in^3 (cubic inch) =	16.39 cc (cubic centimeters)
	16.39 ml (milliliters)
	5.79×10^{-4} ft^3 (cubic feet)
	1.64×10^{-5} m^3 (cubic meters)
	4.33×10^{-3} gal (gallons)
	0.0164 ℓ (liters)

(Continued)

Units of Length

	0.55 fl oz (fluid ounces)
1 m³ (cubic meter) =	1,000,000 cc = 10^6 cc (cubic centimeters)
	35.31 ft³ (cubic feet)
	61,023 in³ (cubic inches)
	264.17 gal (gallons)
	1,000 ℓ (liters)
1 yd³ (cubic yards) =	201.97 gal (gallons)
	764.55 ℓ (liters)
1 gal (gallon) =	3,785 cc (cubic centimeters)
	0.134 ft³ (cubic feet)
	231 in³ (cubic inches)
	3.785 ℓ (liters)
1 ℓ (liter) =	1,000 cc (cubic centimeters)
	1 dm³ (cubic decimeters)
	0.0353 ft³ (cubic feet)
	61.02 in³ (cubic inches)
	0.264 gal (gallons)
	1,000 (milliliters)
	1.057 qts (quarts)
1 qt (quart) =	946.4 cc (cubic centimeters)
	57.75 in³ (cubic inches)
	0.946 ℓ (liters)

Units of Mass

1 gm (grams) =	0.001 kg (kilograms)
	1,000 mg (milligrams)
	1,000,000 mg = 10^6 ng (nanograms)
	2.205×10^{-3} lbs (pounds)
1 kg (kilogram) =	1,000 gms (grams)
	2.205 lbs (pounds)
1 lbs (pound) =	453.59 gms (grams)
	16 oz (ounces)

Units of Time

1 day =	24 hrs (hours)
	1440 min (minutes)
	86,400 sec (seconds)
	0.143 weeks
	2.738×10^{-3} yrs (years)
1 hr (hours) =	0.0417 days
	60 min (minutes)

(*Continued*)

	Units of Length
	3,600 sec (seconds)
	5.95×10^{-3} yrs (years)
1 hr (hour) =	0.0417 days
	60 min (minutes)
	3,600 sec (seconds)
	5.95×10^{-3} weeks
	1.14×10^{-4} yrs (years)
1 min (minutes) =	6.94×10^{-4} days
	0.0167 hrs (hours)
	60 sec (seconds)
	9.92×10^{-5} weeks
	1.90×10^{-6} yrs (years)
1 week =	7 days
	168 hrs (hours)
	10,080 min (minutes)
	6.048×10^{5} sec (seconds)
	0.0192 yrs (years)
1 yr (year) =	365.25 days
	8,766 hrs (hours)
	5.26×10^{5} min (minutes)
	3.16×10^{7} sec (seconds)
	52.18 weeks

UNITS OF THE MEASURE OF TEMPERATURE

°C (degrees Celsius) =	$[(5/9)(°F - 32°)]$
1°C (expressed as an interval) =	$1.8°F = [9/5]°F$ (degrees Fahrenheit)
	1.8°R (degrees Rankine)
	1.0 K (degrees Kelvin)
°F (degree Fahrenheit) =	$[(9/5)(°C) + 32°]$
1°F (expressed as an interval) =	$0.556°C = [5/9]°C$ (degrees Celsius)
	1.0°R (degrees Rankine)
	0.556 K (degrees Kelvin)

UNITS OF FORCE

1 dyne =	1×10^{-5} nt (newtons)
1 nt (newton) =	1×10^{5} dynes

UNITS OF WORK OR ENERGY

1 BTU (British Thermal Unit) =	252 cal (calories)
	1,055.06 j (joules)
	10.41 liter-atmospheres
	0.293 watt-hours
1 cal (calories) =	3.97×10^{-3} BTUs (British Thermal Units)
	4.18 j (joules)
	0.0413 liter-atmospheres
	1.163×10^{-3} watt-hours
1 ev (electron volt) =	1.602×10^{-12} ergs
	1.602×10^{-19} j (joules)
1 erg =	1 dyne-centimeter
1×10^{-7} j (joules)	
	2.78×10^{-11} watt-hours
1 j (joule) =	9.48×10^{-4} BTUs (British Thermal Units)
	0.239 cal (calories)
	10,000,000 ergs = 1×10^7 ergs
	9.87×10^{-3} liter-atmospheres
	1.00 nt-m (newton-meters)
1 kcal (kilocalorie) =	3.97 BTUs (British Thermal Units)
	1,000 cal (calories)
	4,186.8 j (joules)
1 kw-hr (kilowatt-hour) =	3,412.14 BTU (British Thermal Units)
	3.6×10^6 j (joules)
	859.8 kcal (kilocalories)
1 Nt-m (newton-meter) =	1.00 j (joules)
	2.78×10^{-4} watt-hours
1 watt-hour =	3.412 BTUs (British Thermal Units)
	859.8 cal (calories)
	3,600 j (joules)
	35.53 liter-atmospheres

UNITS OF POWER

1 hp (horsepower) =	745.7 j/sec (joules/sec)
1 kw (kilowatt) =	56.87 BTU/min (British Thermal Units/minute)
	1.341 hp (horsepower)
	1,000 j/sec (joules/sec)
1 watt =	3.41 BTU/hr (British Thermal Units/hour)
	1.341×10^{-3} hp (horsepower)
	1.00 j/sec (joules/second)

UNITS OF PRESSURE

1 atm (atmosphere) =	1.013 bars
	10.133 newtons/cm^2 (newtons/square centimeters)
	33.90 ft. of H$_2$O (feet of water)
	101.325 kp (kilopascals)
	14.70 psia (pounds/square inch—absolute)
	760 torr
	760 mm Hg (millimeters of mercury)
1 bar =	0.987 atm (atmospheres)
	1 × 10^6 dynes/cm^2 (dynes/square centimeter)
	33.45 ft of H$_2$O (feet of water)
	1 × 10^5 pascals [nt/m^2] (newtons/square meter)
	750.06 torr
	750.06 mm Hg (millimeters of mercury)
1 inch of water =	1.86 mm Hg (millimeters of mercury)
	249.09 pascals
	0.0361 psi (lbs/in^2)
1 mm Hg (millimeter of merc.) =	1.316 × 10^{-3} atm (atmospheres)
	0.535 in H$_2$O (inches of water)
	1.33 mb (millibars)
	133.32 pascals
	1 torr
	0.0193 psia (pounds/square inch—absolute)
1 pascal =	9.87 × 10^{-6} atm (atmospheres)
	4.015 × 10^{-3} in H$_2$O (inches of water)
	0.01 mb (millibars)
	7.5 × 10^{-3} mm Hg (millimeters of mercury)
1 psi (pounds/square inch) =	0.068 atm (atmospheres)
	27.67 in H$_2$O (inches of water)
	68.85 mb (millibars)
	51.71 mm Hg (millimeters of mercury)
	6,894.76 pascals
1 torr =	1.33 mb (millibars)

UNITS OF VELOCITY OR SPEED

1 fps (feet/second) =	1.097 kmph (kilometers/hour)
	0.305 mps (meters/second)
	0.01136 mph (miles/hours)
1 mps (meters/second) =	196.9 fpm (feet/minute)
	3.6 kmph (kilometers/hour)
	2.237 mph (miles/hour)

(Continued)

1 mph (mile/hour) =	88 fpm (feet/minute)
	1.61 kmph (kilometers/hour)
	0.447 mps (meters/second)

UNITS OF DENSITY

1 gm/cc (grams/cubic cent.) =	62.43 lbs/ft^3 (pounds/cubic foot)
	0.0361 lbs/in^3 (pounds/cubic inch)
	8.345 lbs/gal (pounds/gallon)
1 lbs/ft^3 (pounds/cubic foot) =	16.02 gms/ℓ (grams/liter)
1 lbs/in^2 (pounds/cubic inch) =	27.68 gms/cc (grams/cubic centimeter)
	1.728 lbs/ft^3 (pounds/cubic foot)

UNITS OF CONCENTRATION

| 1 ppm (parts/million-volume) = | 1.00 ml/m^3 (milliliters/cubic meter) |
| 1 ppm (wt) = | 1.00 mg/kg (milligrams/kilograms) |

RADIATION & DOSE RELATED UNITS

1 Bq (becquerel) =	1 radioactive disintegration/second
	2.7×10^{-11} Ci (curie)
	2.7×10^{-8} (millicurie)
1 Ci (curie) =	3.7×10^{10} radioactive disintegration/second
	3.7×10^{10} Bq (becquerel)
	1,000 mCi (millicurie)
1 Gy (gray) =	1 j/kg (joule/kilogram)
	100 rad
	1 Sv (sievert) [unless modified through division by an appropriate factor, such as Q and/or N]
1 mCi (millicurie) =	0.001 Ci (curie)
	3.7×10^{10} radioactive disintegrations/second
	3.7×10^{10} Bq (becquerel)
1 rad =	100 ergs/gm (ergs/gm)
	0.01 Gy (gray)
	1 rem [unless modified through division by an appropriate factor, such as Q and/or N]
	1 rem = 1 rad [unless modified through division by an appropriate factor, such as Q and/or N]
	1 Sv (sievert) = 1 Gy (gray) [unless modified through division by an appropriate factor, such as Q and/or N]

DID YOU KNOW?

Units and dimensions are not the same concepts. Dimensions are concepts like time, mass, length, weight, etc. Units are specific cases of dimensions, like hour, gram, meter, lb, etc. You can multiply and divide quantities with different units: 4 ft × 8 lb = 32 ft-lb; but you can add and subtract terms only if they have the same units: 5 lb + 8 kg = **NO WAY!!!**

Geologic Time Scale

Erathem or Era Epoch	System, Subsystem or Period, Subperiod	Series or Epoch
Cenozoic 65 million years ago to present "Age of Recent Life"	**Quaternary** 1.8 million years ago to the Present	**Holocene** 11,477 years ago (+/-85 years) to the Present—Greek "holos" (entire) and "ceno" (new)
		Pleistocene 1.8 million to approx. 11,477 (+/- 85 years) years ago—The Great Ice Age—Greek Words "pleistos" (most) and "ceno" (new).
	Tertiary 65.5 to 1.8 million years ago	**Pliocene** 5.3 to 1.8 million years ago—Greek "pleion" (more) and"ceno" (new).
		Miocene 23.0 to 5.3 million years ago— Greek"meion" (less) and"ceno" (new).
		Oligocene 33.9 to 23.0 million Years ago—Greek "oligos" (little, few) and "ceno" (new).
		Eocene 55.8 to 33.9 million years ago—Greek "eos" (dawn) and "ceno" (new).
		Paleocene 65.5 to 58.8 million years ago—Greek "palaois" (old) and "ceno" (new)
Mesozoic 251.0 to 65.5 million years ago—Greek means "middle life"	**Cretaceous** 145.5 to 65.5 million years ago "The Age of Dinosaurs"	Late or Upper Early or Lower
	Jurassic 199.6 to 145.5 million years ago	Late or Upper Middle Early or Lower
	Triassic 251.0 in 199.6 million years ago	Late or Upper Middle Early or Lower
Paleozoic 542.0 to 251.0 million years ago, "Age of Ancient Life"	**Permian** 299.0 to 251.0 million years ago	Lopingian Guadalupian Cisuralian
	Pennsylvanian 318.1 to 299.0 million years ago "The Coal Age"	Late or Upper Middle Early or Lower

(Continued)

Erathem or Era Epoch	System, Subsystem or Period, Subperiod	Series or Epoch
	Mississippian 359.2 to 318.1 million years ago	Late or Upper Middle Early or Lower
	Devonian 416.0 to 359.2 million years ago	Late or Upper Middle Early or Lower
	Silurian 443.7 to 416.0 million years ago	Pridoli Ludlow Wenlock Llandovery
	Ordovician 488.3 to 443.7 million years ago	Late or Upper Middle Early or Lower
	Cambrian 542.0 to 488.3 million years ago	Late or Upper Middle Early or Lower

Precambrian
approximately 4 billion years ago to 542.0 million years ago

Prologue

IS IT THE BEGINNING OF THE END OR THE BEGINNING OF THE BEGINNING?

When we consider that we are probably at or near peak oil natural production and already absorbing the higher costs associated with it (i.e., if we are able to obtain a reliable and consistent supply of it in the first place), renewable energy alternatives look a lot more viable and necessary to us. On the other hand, currently (2023), many feel that we are suffering severe economic conditions, and if we are lucky enough to have reliable and consistent employment with accompanying benefits, the advantages of switching from oil/natural gas supplies to renewable energy sources may not be that apparent or pressing to a fortunate but misinformed, mislead, and misguided few. Until the summer of 2008, when in July gasoline prices at the pump were $4.00+, many of us did not give much thought to peak oil/natural gas or possible alternatives. After the spike in gasoline prices, however, people (those with and without employment) started to look at hybrid cars and domestic renewable energy alternatives in a very new light. Using the east and west stretch of Interstate 40 as our example, consider that when driving on it, it is not that difficult or unusual to pick out numerous hybrid automobiles along with those ubiquitous heavy trucks, RVs, and other standard vehicles sharing the roadway. In addition, at various locations along the interstate it is not unusual to find giant wind towers, their turbine blades slashing through the air turning turbine-generator shafts cranking out electricity. At various locations along that same super-highway, especially at various rest stops (i.e., the ones not closed due to budget cuts) and state visitor centers, many have small arrays of solar cells contained in individual panels positioned here and there. These remotely stationed solar panels supply electricity to fans, signal lights, natural gas metering systems, alarm systems, pump monitoring equipment, digital signs, and many other small appliances and industrial devices.

It is not unusual for drivers and passengers in those vehicles traveling the interstates, when they see those conspicuous wind turbines, solar panels, and hydroelectric dams, not to be that impressed with these generators of renewable energy. This is usually the case, of course, because their concern is focused on getting from point A to point B and maybe safely back to point A. These folks may miss the main point, however. The main point is that someday we will have (hopefully) developed high-capacity, highly efficient electric or hydrogen-like fuel-cell vehicles that will be as common as the gas- and diesel-powered vehicles of today. Much of the electricity that will be needed to recharge electric vehicles or produce fuel-cell-powered vehicles will be provided by generators of renewable electricity including those same wind turbines, solar panels, and hydroelectric dams.

In regard to the present economic situation driven by many factors, including the current high price of crude oil and its dwindling accessibility, we ask ourselves this question: Is it the beginning of the end or the beginning of the beginning? As we send this manuscript to the publisher for production, we ask ourselves this question and several others. So, what do we mean? What are we talking about? Is doomsday around the corner? Is doomsday here already and we are too ignorant to realize it, or just don't give a damn, or are we mesmerized by convoluted and confused happenings clouding our ability to recognize the signs—the beginning of the end or the beginning of the beginning—do we anticipate the end, or an end—whatever that end might be?

The end of what you might ask?—Geez, we hope that is what you are asking yourselves and us. We mean the end of the so-called good life as we know it… that is, that good life that our parents (the greatest generation), and we, the so-called grasshopper generation, knew very briefly, yesterday, in the not-so-distant past…or, as we can only hope, as we will know again tomorrow.

As "we knew before yesterday" refers to the era just before that infamous dot. com bust in 1999, or thereabouts. Remember those days when young computer whiz kids were driving around in all those BMWs and living the good life in million-dollar homes; all mortgaged to the hilt, of course, with portfolios filled with promissory notes? But the mantra then was not to worry; dot.com wizards were ultra-millionaires—on paper at least. They were the new rich; that is, until the bust. When that occurred their paper was just paper and almost worthless, certainly not as utilitarian as toilet paper; just paper with print that promised untold fortunes—that same print is now metaphorically invisible, of course. Then there are those recent years when those dysfunctional government regulators, those asleep at the switch responsible-principals-in-charge, and uncaring, misguided, unguided, and totally dysfunctional congressional leaders allowed just about anyone and everyone to buy a home—even those with bad credit or no credit or no job or no future—those non-qualified buyers who brought us to the sharp edge of the abyss we are now straddling…and if not careful working to push us over.

So, is it all doom and gloom? Short answer: We do not have a clue. Long answer: not necessarily if we make the right moves soon. Right moves? What are they? How do we maintain the good life? How do we improve on our lives to the point that we are living as well as or better than our parents did?

We do not profess to know all the problems facing our country at present nor do we know all the solutions. We know what we know from what we know—from what we see and feel. Our great country is at a crossroads. We can maintain our present lifestyles; fall on our swords; or progress through innovation. We hope it is progression, moving forward in maintaining the good life that is before us. This is our hope.

So, how do we progress—how do we maintain that so-called good life? In our opinions, the first step is we must obtain energy self-sufficiency, independence, innovation, a fundamental shift from fossil fuels to renewable, non-polluting energy supplies. Energy is the lifeblood of America. We must get off our knees and stop begging those who hate us to supply us with the energy we need. We must put America first. Using American genius we must innovate and come up with a solution to the pending energy crisis. Remember, every problem has a solution; we must find the solution to the energy crisis. Moreover, when we do innovate, we must make sure that America holds onto its trade secrets, on how to produce renewable energy sources; otherwise, various countries will simply steal our ideas and manufacture energy at rates for which we can't compete. That is, for every $9.00 of production cost (mainly for labor, raw materials, and taxes) we spend in the United States, many foreign countries simply provide the same service and build the same product, but for $1.00 or less. If we are going to innovate, let's innovate for America. We have given away too much, and now we are paying for it at the pump and elsewhere.

Is lithium really the new oil?

Maybe this book will answer this question—and others.

About the Author

Frank R. Spellman is a retired assistant professor of Environmental Health at Old Dominion University, Norfolk, VA and author of more than 155 books. Spellman has been cited in more than 400 publications. He serves as a professional expert witness and incident/accident investigator for the U.S. Department of Justice and a private law firm and consults on homeland security vulnerability assessments (VAs) for critical infrastructure including water/wastewater facilities nationwide. Dr. Spellman lectures on sewage treatment, water treatment, and homeland security and health and safety topics throughout the country and teaches water/wastewater operator short courses at Virginia Tech (Blacksburg, VA). He holds a BA in Public Administration; BS in Business Management; MBA, Master of Science, MS, in Environmental Engineering and PhD Environmental Engineering.

Part 1

Introduction

1 Lithium, What Is It?

THIRD PLACE ELEMENT

This book properly begins by starting with an explanation of the periodic table. In Figure 1.1, the periodic table is shown and for those of you who are chemists or other scientists you are very familiar with the table. However, for those who are not that familiar with the table, a brief introduction is called for here and is provided. Note that lithium is listed as 3 of 118 elements on the periodic table.

PERIODIC TABLE

Lithium (Li) is listed and shown in the periodic table (see Figure 1.1) in the six alkali metals—lithium (chemical symbol: Li), sodium (Na), potassium (K), rubidium (Rb), cesium (Cs), and francium (Fr). These are all important chemicals and without sodium and potassium we could not live. With the exception of lithium, which is very

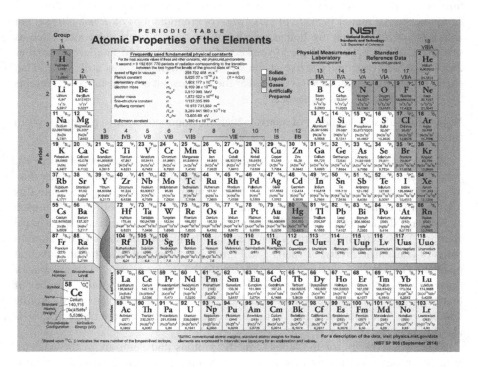

FIGURE 1.1 The periodic table. Lithium (Li) can be found on the far left-hand side of the periodic table in group 1 or 1 A in the alkali metals directly beneath hydrogen (H).

Source: NIST Periodic Table.

DOI: 10.1201/9781003387879-2

common, rubidium, cesium, and francium are rare. Lithium again is not rare and is used extensively in the batteries that power up small electronic devices including laptop computers, MP3 players, pocket calculators, and several other devices. Lithium is also used as the critical element in electric vehicle (EV) car batteries.

Lithium along with the other alkali metals are found on the far-left side of the periodic table in group 1 or 1A as shown in Figure 1.1. The periodic table is arranged according to increasing atomic number. The rows on the periodic table are called periods. Lithium is in period 2. The columns of the periodic table are called groups or families. Lithium along with the other alkali metals are in group 1.

Note: To gain understanding of the makeup of Li and its electrical properties, it is important to have a basic understanding of matter. The following explanation of matter is given in the sense of electron flow (aka electricity) because it is the electrical aspects of lithium that is the focus of this text. However, the current uses of lithium, including medical uses, are discussed in this text.

STRUCTURE OF MATTER

Matter is anything that has mass and occupies space. To study the fundamental structure or composition of any type of matter, it must be reduced to its fundamental components. All matter is made of *molecules*, or combinations of *atoms* (Greek: not able to be divided), that are bound together to produce a given substance, such as salt, glass, or water. For example, if you keep dividing water into smaller and smaller drops, you will eventually arrive at the smallest particle that was still water. That particle is the molecule, which is defined as the smallest bit of a substance that retains the characteristics of that substance.

Note: Molecules are made up of atoms, which are bound together to produce a given substance.

Atoms are composed, in various combinations, of subatomic particles of *electrons, protons,* and *neutrons*. These particles differ in weight (a proton is much heavier than the electron) and charge. We are not concerned with the weights of particles in this text, but the *charge* is extremely important in electricity and in lithium when used as a storage device (battery) of electricity. The electron is the fundamental negative charge (−) of electricity. Electrons revolve about the nucleus or center of the atom in paths of concentric *orbits*, or shells (see Figure 1.2). The proton is the fundamental positive (+) charge of electricity. Protons are found in the nucleus. The number of protons within the nucleus of any particular atom specifies the atomic number of that atom. For example, the helium atom has two protons in its nucleus so the atomic number is 2. The neutron, which is the fundamental neutral charge of electricity, is also found in the nucleus.

Most of the weight of the atom is in the protons and neutrons of the nucleus. Whirling around the nucleus are one or more negatively charged electrons. Normally, there is one proton for each electron in the entire atom so that the net positive charge of the nucleus is balanced by the net negative charge of the electrons rotating around the nucleus (see Figure 1.3).

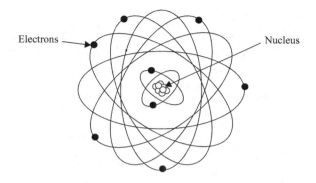

FIGURE 1.2 Electrons and nucleus of an atom.

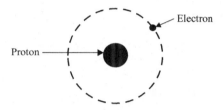

FIGURE 1.3 One proton and one electron = electrically neutral.

Note: Most batteries are marked with the symbols + and − or even with the abbreviations POS (positive) and NEG (negative). The concept of a positive or negative polarity and its importance in lithium-stored electricity will become clear later. However, for the moment, you need to remember that an electron has a negative charge and that a proton has a positive charge.

Earlier, it was pointed out that in an atom the number of protons is usually the same as the number of electrons. This is an important point because this relationship determines the kind of element (the atom is the smallest particle that makes up an element; an element retains its characteristics when subdivided into atoms) in question. Figure 1.4 shows a simplified drawing of several atoms of different materials based on the conception of electrons orbiting about the nucleus and the atomic configuration of lithium. For example, hydrogen has a nucleus consisting of one proton, around which rotates one electron. The helium atom has a nucleus containing two protons and two neutrons with two electrons encircling the nucleus. Both of these elements are electrically neutral (or balanced) because each has an equal number of electrons and protons. Since the negative (−) charge of each electron is equal in magnitude to the positive (+) charge of each proton, the two opposite charges cancel.

A balanced (neutral or stable) atom has a certain amount of energy, which is equal to the sum of the energies of its electrons. Electrons, in turn, have different energies called *energy levels*. The energy level of an electron is proportional to its distance from the nucleus. Therefore, the energy levels of electrons in shells farther from the nucleus are higher than that of electrons in shells nearer the nucleus.

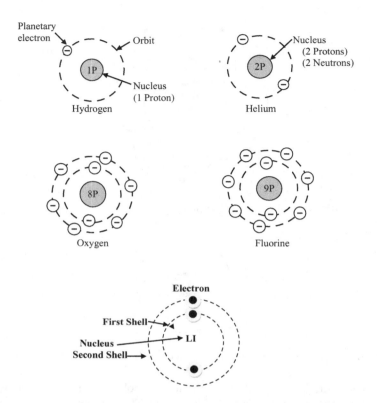

FIGURE 1.4 Atomic structure of a few elements and atomic configuration of Li ([He]2s1).

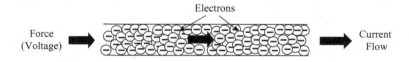

FIGURE 1.5 Electron flow in a copper wire.

When an electric force is applied to a conducting medium, such as copper wire, electrons in the outer orbits of the copper atoms are forced out of orbit (i.e., liberating or freeing electrons) and are impelled along the wire. This electrical force, which forces electrons out of orbit, can be produced in a number of ways, such as by moving a conductor through a magnetic field; by friction, as when a glass rod is rubbed with cloth (silk); or by chemical action, as in a battery.

When the electrons are forced from their orbits, they are called **free electrons**. Some of the electrons of certain metallic atoms are so loosely bound to the nucleus that they are relatively free to move from atom to atom. These free electrons constitute the flow of an electric current in electrical conductors.

Note: When an electric force is applied to a copper wire, free electrons are displaced from the copper atoms and move along the wire, producing electric current flow as shown in Figure 1.5.

If the internal energy of an atom is raised above its normal state, the atom is said to be **excited**. Excitation may be produced by causing the atoms to collide with particles that are impelled by an electric force as shown in Figure 1.5. In effect, what occurs is that energy is transferred from the electric source to the atom. The excess energy absorbed by an atom may become sufficient to cause loosely bound outer electrons (as shown in Figure 1.5) to leave the atom against the force that acts to hold them within.

Note: An atom that has lost or gained one or more electrons is said to be **ionized**. If the atom loses electrons, it becomes positively charged and is referred to as a **Positive Ion**. Conversely, if the atom gains electrons, it becomes negatively charged and is referred to as a **Negative Ion**.

ATOMIC STRUCTURE OF LITHIUM

A neutral lithium atom has three protons and three electrons (number 3 of 118 elements on the periodic table; Figure 1.6 shows sample of lithium in oil and lithium's

FIGURE 1.6 Lithium samples stored in oil (for safety) and lithium's chemical symbol indicating atomic number 3 and approximate atomic weight of 7.

chemical symbol). On the periodic table, an element's period, or row, indicates how many energy levels are occupied by the electrons in its atoms. Lithium is in period 2, indicating the element's three electrons are found on two energy levels. Two electrons are held in the first energy level. The last electron is found on the second, outermost energy level. Note that electrons found on the outermost energy level of an atom are called *valence* electrons. Valence electrons are important because they are the electrons that are involved in chemical bonding with other atoms.

BASIC PROPERTIES OF LITHIUM

A few important properties of lithium are given in Table 1.1.

TABLE 1.1
Properties of Lithium

Property	Value
Atomic mass	6.941 gm
Density	0.535 gm/cm^3
Atomic number	3
Atomic symbol	Li
Appearance	Silvery white
Atomic radiance	152 pm
Melting point	180.5°C (356.9°F)
Boiling point	1,342°C (2,448°F)
Critical point	3,220 K, 67 MPa
Heat of fusion	3.00 kJ/mol
Heat of vaporization	136 kJ/mol
Molar heat capacity	24.860 j/(mol·K)
Oxidation state	+1
Electronegativity[a]	0.98 (Pauling Scale)
Electronic configuration	[He]2s1

[a] *Electronegativity, symbolized Y, is the tendency for an atom of a given chemical element to attract shared electrons while forming a chemical bond. As shown in Table 1.1, lithium's electronegativity is 0.98 on the Pauling Scale. This value, 0.98, is based on calculation because electronegativity can't be measured it must be calculated.*

LITHIUM COMPOUNDS

Many different lithium salts can be used as medication, including lithium carbonate, lithium acetate, lithium sulfate, lithium citrate, lithium orotate, and lithium gluconate. Note that lithium in pure form has limited but important uses, but lithium compounds such as lithium carbonate, lithium stearate, and lithium hydroxide are widely used in everyday life and for our purposes are briefly described in the following.

LITHIUM CARBONATE

Lithium carbonate (Li_2CO_3) only found in the anhydrous state, is the lithium white salt of carbonate and is an inorganic compound widely used in the processing of metal oxides. Lithium carbonate is extracted primarily from spodumene in pegmatite deposits and lithium salts from underground brine reservoirs. Lithium carbonate is added to glass or ceramic to make the materials stronger. Pyrex™ glass cookware is a good example of where lithium carbonate is used. It is also used in mirrors and lenses for telescopes. Lithium carbonate is also used to extract aluminum metal from aluminum ores. The main use of lithium carbonate is in lithium-ion batteries where its compounds are used as the cathode and the electrolyte. As early as the 1840s, lithium carbonate was used to treat stones in the bladder, and by the 1850s, lithium carbonate salts were used to treat grout, urinary calculi, rheumatism, mania, depression, and headache. By the 1940s, lithium carbonate became a common treatment for bipolar disorder.

LITHIUM STEARATE

Lithium stearate ($LiC_{18}H_{35}O_3$) is a white solid powder that is mixed with petroleum to make a heavy, lubricated grease used in many industrial machines; it is also used in making soaps. The grease does not break down at high temperatures and does not get hard at low temperatures. It also does not react with oxygen in air or in water, so the chemical composition has a long use life.

LITHIUM HYDROXIDE

Lithium hydroxide (LiOH) is an organic compound that can exist as anhydrous or hydrated and that is useful in absorbing carbon dioxide.

THE NEW OIL

Because demand is soaring lithium is often referred to as "the new oil" and also as "the new white gold" because it is seen as a valuable alkali metal that is able to store huge amounts of energy all of it squeezed into a very small area, space, and/or container—such as a storage battery. The demand for lithium is soaring and competition to find and produce lithium is significant and in some cases it is filled with drama.

REFERENCE

NIST Periodic Table. (2020). Accessed 10/17/2022 @ http://www.NIST.gov. (U.S. Department of Commerce).

2 Manifestation

The manifestation or occurrence of lithium is three-fold from astronomical, terrestrial, and biological.

ASTRONOMICAL MANIFESTATION: CLASSICAL NOVAE

NASA (2020) in a funded study suggests "that most of the lithium in our solar systems—and even in the galaxy—came from bright stellar explosions called classical novae." When a white dwarf—a remnant about the diameter of the Earth and with the mass of the Sun—is orbited by a larger star, a classical nova occurs. The larger star emits gas onto the white dwarf and when enough gas has accumulated on the white dwarf, a nova (or explosion) occurs. About 50 of these explosions occur each year in our galaxy, and the brightest ones are observed worldwide by astronomers (NASA, 2020). NASA researchers have determined that a small amount of lithium was created in the formation of the universe. The majority of lithium is manufactured in the nuclear reactions that power the nova explosions. NASA researchers suggest that the nova explosions would then distribute that lithium throughout the galaxy and deliver most of the lithium we use currently in electronics and medicine (NASA, 2020).

TERRESTRIAL MANIFESTATION

Deep below the Salton Sea in California is a treasure trove of lithium. You know, the element used in manufacturing the best batteries available at the present time.

FIGURE 2.1   Salton Sea, California.

Source: USGS Public Domain Photo (2019).

DOI: 10.1201/9781003387879-3

FIGURE 2.2   Spodumene. Photo by Frank R. Spellman.

Mining for lithium can be considered the "new oil" or a modern gold rush, of sorts. The "oil or gold" is the white mineral lithium that is removed from the geothermal brine at the Salton Sea (see Figure 2.1). Now it is important to point out, even though it has been said that the salt of the Salton Sea turns to gold this is not quite true—the lithium does not come from the salt in the Salton Sea, this is simply a misconception. Instead, the lithium comes from the superheated fluid (brine) deep below and the extraction procedure is a geothermal process.

The two main sources of lithium for commercial purposes are spodumene mines and salar brine water from Salar de Atacama in northern Chile.

SPODUMENE

Lithium mining of spodumene from the Earth is accomplished using traditional mining processes. Spodumene, hard rocks and clusters of crystals (see Figure 2.2), is a pyroxene mineral consisting of lithium aluminum inosilicate, $LiAl(SiO_3)^2$ and this is where the lithium manifests from. Spodumene is found worldwide, and at the present time, the largest spodumene mine is located in Australia.

SALAR BRINE WATER

Over half the world's known lithium are contained in salar brine water sources. Most of the proven lithium-containing salars are found in South America with most in Chile and some in Bolivia and Argentina. Note that although salar brine water mining is being conducted in these South American countries, there has been political and environmental interruptions.

BIOLOGICAL MANIFESTATION

Lithium is found in trace amount in many plants, plankton, and invertebrates, but only in concentrations in parts per billion (ppb).

REFERENCES

NASA (2020). Lithium comes from exploding stars. Accessed 10/24/22 @ https://www.nasa.gov/feature/lithium-comes-from-exploding-stars.

USGS Public Domain Photo (2019) Accessed 12/7/2021 @ https://www.usgs.gov.media/images/saltonsea.

3 Production

INTRODUCTION

The new oil, lithium is the lightest of all metals and in this book 19 uses are identified and discussed including its use in air treatment, in construction of batteries, ceramics, glass, metallurgy, pharmaceuticals, and polymers. Currently, the big push, so to speak, in producing lithium is for its rechargeable lithium-ion batteries and the movement toward saving the environment by reducing the global climate change that is partly attributable to the burning of fossil fuels. As pointed out earlier, today's practice is to extract lithium from brines that are pumped from beneath arid sedimentary basins and also extracted from granitic pegmatite ores. Pegmatite is an igneous rock showing a very coarse texture, with large interlocking crystals. The leading producer of lithium from pegmatites is Australia. Other sources of lithium include clays.

In regard to lithium in clays, it is widespread in clay minerals in small amounts. When present in clays, the lithium may be an impurity, an inclusion, inside lattice cavities, adsorbed on the surface, or by isomorphous substitution (i.e., the replacement of one atom by another of similar size in a crystal structure without disrupting or seriously changing the structure). Isomorphous substitution is the most common occurrence. The clay containing the largest amount of lithium is swinefordite which is mainly found at the Foote Lithium Co. mine at Kings Mountain in North Carolina. Worldwide resources of lithium are estimated to be more than 39 million metric tons but note that the United States is not a producer at present—but has significant lithium resources (USGS, 2017).

MINERALOGY AND DEPOSIT TYPES

Lithium is classed as a large-ion lithophile element. It is interesting to note that in addition to lithium being a lithophile element, this is the case with rare earth elements (REEs). What lithophile means is like REEs lithophile in the melt remain on or close to the surface because they combine with oxygen, forming compounds that do not sink into the Earth's core. Over the ages concentrated lithium in the continental crust came about via active plate tectonics which caused partial melting of the mantle beneath mid-ocean ridges and volcanic arcs (i.e., in a chain of volcanoes). What happens is that the melt, or magma, rises to the surface and cools to become new rock in Earth's crust, bringing much of the available lithium with it. Among the common rock or sediment types, the highest lithium concentration are in shales (~66 ppm), deep sea clays (~57 ppm), and low-calcium granites (~40 ppm) (Faure, 1998). These trace concentrates are not sufficient for an ore deposit or even for the formation of minerals in which lithium is part of the chemical formula. In line with being present only in trace concentrations, lithium atoms substitute for other metals—typically magnesium—in common rock-forming minerals (USGS, 2017).

DOI: 10.1201/9781003387879-4

In order to lithium to form rare combinations of rare factors must be present. Currently, most of the known lithium minerals are found in coarsely crystalline granites known as lithium-cesium-tantalum (LCT) pegmatites. With regard to mineral resources, the most important minerals are the potassium lithium aluminum silicates spodumene and petalite and the potassium lithium aluminum silicate pink mica lepidolite. In sedimentary rocks, the main lithium mineral is the clay hectorite (a rare greasy white clay that may contain up to 1%–1.2% lithium. Additional information on lithium minerals is provided in Table 3.1.

During weathering of rocks, lithium because it is extremely soluble it tends to be removed in solution and carried to the sea by rivers. Thus, you might expect to have lithium build in the oceans in the same way that a buildup of sodium has made the oceans salty. However, when sampled seawater contains less than 1-ppm lithium. So, the question becomes where is all the lithium that the rivers have conveyed to the oceans? The most likely explanation is that the lithium accumulates in seafloor oozes and resides in clay minerals.

With regard to lithium deposit types, they are listed as follows: closed-basin brines, 58%; pegmatites (including lithium-enriched granites), 26%; lithium clays (hectorite), 7%; and oilfield brines, geothermal brines, and lithium-zeolites (jadarite), 3% each (Evans, 2012).

TABLE 3.1
Commercially and/or Scientifically Important Lithium-Bearing Minerals

Mineral Name	Chemical Formula	Lithium	Geologic Setting (weight %)
Amblygonite	$(Li, Ba)Al(PO_4)(F, OH)$	3.44	Pegmatite intrusions in orogenic belts
Elbaite	$Na(Li, Al)_3Al_6(BO_3)_3Si_6O_{18}(OH)_4$	1.89	Tourmaline group. Pegmatite intrusions in organic belts. Includes the form watermelon tourmaline
Eucryptite	$LiAlSiO_4$	5.51	Pegmatite intrusions in orogenic belts
Hectorite	$Na_{0.3}(Mg, Li)_3Si_4O_{10}(OH)_2$	0.54	Hydrothermal alteration of volcanic ash in arid, closed basins
Jadarite	$LiNaB_3SiO_7(OH)$	3.38	Hydrothermal alteration of volcanic ash in arid, closed basins
Lepidolite	$(Li, Al)_3(Si, Al)_4O_{10}(F, OH)_2$	3.58	Pegmatite intrusions in orogenic belts
Montrebasite	$LiAl(PO_4)(OH, F)$	4.74	Pegmatite intrusions in orogenic belts
Petalite	$LiAlSi_4O_{16}$	2.09	Pegmatite intrusions in orogenic belts
Spodumene	$LiAlSi_2O_6$	3.73	Pegmatite intrusions in orogenic belts. Gem forms are triphane (yellow), Kunzite (pink), and hiddenite (green)
Zabuyelite	Li_2CO_3	18.79	Evaporite mineral (rare)

Source: Based on chemical compositions adapted from Barthelmy (2014).

LCT PEGMATITE DEPOSITS

Pegmatites are classified (for our purposes) as common pegmatites which have the simple mineralogy of granites. The second classification is the rare pegmatites that consist of two families, the LCT pegmatites and the niobium, yttrium, and fluorine NYF pegmatites.

So, we now know the two classifications of pegmatites that are the subject of this book.

So, what are pegmatites? Well, many experts accept that pegmatites are an essentially igneous rocks, commonly of granitic composition, that is distinguished from other igneous rocks by its extremely coarse but variable grain size, or by an abundance of crystals with skeletal, graphic, or other strongly directional growth habits. LCT pegmatites are a petrogenetically (i.e., creation of new rock) defined subset of granitic pegmatites that are associated with certain granites. Their composition is mostly of quartz, potassium feldspar, albite, and muscovite—common accessory minerals include garnet, tourmaline, and apatite. In this book, we are concerned about lithium in particular and the major lithium ore minerals are spodumene, petalite, and lepidolite; cesium mainly comes from pollucite; and tantalum mostly comes from columbite-tantalite (USGS, 2017). Also occurring in LCT pegmatites are tine ore, cassiterite, and beryllium ore, beryl. Note that a number of gemstones and high-value museum specimens of rare minerals also occur in the LCT pegmatites. Among the gemstones are the beryl varieties emerald, heliodor, and aquamarine; the spodumene varieties kunzite and hiddenite; and watermelon tourmaline. LCT pegmatites are also mined for ultrapure quartz, potassium feldspar, albite, and muscovite (Bradley and McCauley, 2016).

Australia, Brazil, China, Portugal, and Zimbabwe are the main producers of lithium from pegmatites. The world has hundreds of known deposits, and some of these contain large reserves of lithium that have not been mined; Quebec, Canada, is a good example. Currently, the United States is not producing lithium from pegmatites even though large reserves remain in the Kings Mountain pegmatite district in North Carolina. Other domestic deposits do not meet the economics requirements to be viable because the deposits are too small under standard market conditions.

It is interesting to note that the three countries Chile, Bolivia, and Argentina make up a region known as the Lithium Triangle. It is the high-quality salt flats that the Lithium Triangle is known for. These salt flats include Bolivia's Salar de Uyuni, Chile's Salar de Atacama, and Argentina's Salar de Arizaro and contain 75% of existing known lithium reserves (Halpern, 2014).

LITHIUM-ENRICHED GRANITES

The elements lithium, tantalum, tin, and fluorine are included in zones in some muscovite-bearing granites. In China, at the Yichun Mine in Jiangxi Provinces, the top of a biotite-muscovite granite grades into muscovite granite and then into lepidolite granite, which has been mined for lithium and tantalum (Schwartz, 1992). LCT pegmatites are closely related to lithium-enriched granites, and the two have not been distinguished from one another in recent global assessments of lithium resources (e.g., Kunasz, 2006; Evans, 2008; Yaksic and Tilton, 2009),

CLOSED-BASIN LITHIUM BRINE DEPOSITS

Lithium brine deposits are accumulations of saline groundwater that are enriched in dissolved lithium. Evans (2012) estimates that closed-basin brine deposits contain an estimated 58% of the world's identified lithium resources. When I investigated the basin lithium brine deposits I found that the actual producing deposits have average lithium concentrations that range from 160 to 1,400 ppm and estimated resources of 0.3–6.3 million metric tons of lithium. The producing deposits are located in Asia, North America, and South America, and lie within the northern and arid latitudinal belts on either side of the equator. These deposits share a number of characteristics, including the following (USGS, 2017):

- an arid climate
- closed basin containing a salt lake or salt flat
- tectonically driven subsidence
- associated igneous or geothermal activity
- lithium-bearing source rocks
- one or more adequate aquifers, to host the brine reservoir
- sufficient time to concentrate a brine.

At the present time, one of the best ways to highlight some of the key characteristics of closed-basin lithium brine deposits is to describe the lithium brine deposit in Clayton Valley, Nevada. Lithium has been produced from brine in Clayton Valley since 1966. There is nothing special or even notable about the basin—it has a relatively small footprint of about 100 km² and a total lithium resource of about 0.3 million metric tons—but this site is better understood via experience, actual practice, and observation (USGS, 2017).

The interaction of plates in the Pacific Ocean formed Clayton Valley and about 150 other basins in the Basin and Range Provinces of the Western United States. In the past 15 million years, Nevada has been stretched in an east-west direction to about twice its former width. The consequence of this is the formation of the basins which are lined with faults. Along these north-south trending fault mountains were uplifted and valleys were formed producing the distinctive alternating pattern of linear mountain ranges and valleys. These are relatively flat regions where Earth's former surface has subsided over geologic time, and the resulting basins have been filled with hundredths and thousands meters thick of sediment (USGS, 2017).

DID YOU KNOW?

In its pure elemental form, lithium is soft, slivery-white metal but is highly reactive and therefore never is found as a metal in nature.

It is assumed that the lithium in Clayton Valley came from various sources; some of it was weathered from rocks or sediments by rain, snowmelt, or groundwater, and some of it was brought in by hydrothermal waters that rose from irregular hot rocks below the basin. Note that these hydrothermal waters came to the surface until interrupted by over-pumped groundwater.

Also important to production is climate, an arid climate and its solar evaporation to extract lithium. Garrett (2004) reports that in Clayton Valley, brine carrying an average of 160-ppm lithium is pumped to the surface and evaporated in succession of nine artificial ponds that average a total area of 16 km². In each pond in the chain, brine enters at one end, loses some of its water during the ensuing weeks or months, and is drained of pumped from the other end into the next pond. Similar to the chain of unit processes used in water and wastewater treatment plants multiple ponds are used so as to keep separate the various evaporite minerals that crystallize in sequence. Note that it takes almost 2 years to achieve the "mother liquor;" that is, achieving concentrate that is both enriched in lithium (~5,000 ppm) and depleted in other, more abundant, elements. The last pond pumps the "mother liquor" to a chemical plant for production of lithium carbonate and lithium hydroxide.

The most notable of several lithium brine systems in China is at Zephyr Lake, Tibet. It is the only known closed basin in which a lithium salt (specifically, Zabuyelite) precipitates as part of the evaporite mineral sequence, and the only basin in a recent global compilation where volcanism or hydrothermal activity has not been stated. When measured, the brines had an average concentration of 700 ppm and an estimated resource of about 1.5 million metric tons of lithium (Gruber et al., 2011).

In addition to brines containing lithium in closed basins, there are other brines that contain lithium. Deep oilfield brines, for example, may contain up to several hundreds of parts per million lithium. Evans (2008) reported high tonnages of lithium contained in oilfield brines in Arkansas, North Dakota, Oklahoma, Texas, and Wyoming, with lithium concentrations up to 700 ppm. In some places, the brine contains up to 692 milligrams per liter (mg/L) lithium. Garrett (2004) points out that the brine occupies pore space in an approximately 200-m-thick limestone at depths of 1,800–4,800 m. The brine has been interpreted to be trapped seawater that was subsequently hydrothermally enriched in lithium and other trace elements.

As potential sources of lithium, oilfield brines have two problems. First, they typically occur at much greater depths (greater than 1 km) than closed-basin brines. Second, unless they happen to be located in an arid climate, recovery of lithium using the expedient and inexpensive method of solar evaporation will not be feasible (Bradley and McCauley, 2010).

Another potential source of lithium are the geothermal brines. Note that the geothermal fields around the world have produced geothermal brine and injected the brine back underground but now it's clear that the brines produced at the Salton Sea geothermal filed contain an immense amount of lithium. Geothermal energy comes from hot salty water, or geothermal brine, is pumped to the surface and converted to a gas that turns a turbine to generate electricity from heat within the Earth. In addition to electricity production, these geothermal brines can yield lithium, brought up in the brine solution from thousands of feet underground. Reportedly, lithium is now being recovered by Simbol, Inc. from a geothermal brine in the Salton Sea (see Figure 2.1) along the California-Mexico border (Simbol, Inc., 2013).

Note that removing lithium from geothermal brine is no picnic—very difficult and costly to accomplish. Think about an analogy by Julie Chao (2021) provided herein about the difficulty of extracting any mineral from geothermal brine.

> If you had a jar of marbles of many different colors but wanted only the green ones, how could you efficiently pick them out? What if it wasn't marbles but a jar of glitter, and there was sand, glue, and mud missed in? That begins to describe the complexity of the brine pumped out from beneath California's Salton Sea as part of the geothermal energy production.

LITHIUM CLAY DEPOSITS

A small subset of the world's clay deposits is enriched in lithium. Evans (2012) reports that lithium-bearing clay deposits contain an estimated 7% of the world's lithium resources. Because of hydrothermal processes, a combination of magnesium, fluorine, and lithium in clays has resulted in several locations in the Western United States (Stillings and Morissette, 2012).

Hectorite is a member of the smectite family of clay minerals. The trioctahedral smectites contain the largest amounts of lithium; dioctahedral smectites contain little lithium.

The Kings Valley hectorite deposit, located within the McDermitt caldera complex, and extinct volcanic center in northern Nevada is the location where the most significant lithium-clay resources are found. During the Miocene, the McDermitt area had extensive volcanism; at least five collapsed vents and resurgent domes have been recognized within the complex. The lithium clays occur in hydrothermally altered, volcanic-derived sediments of lakes that occupied the caldera. A proven feasible method of lithium recovery is by leaching the clay with sulfuric acid (Eggleston and Hertel, 2008).

LITHIUM-ZEOLITE DEPOSITS

The Neogene basin system in the Balkan region of Eastern Europe is the only documented lithium-zeolite deposit. Miocene-age beds in the Jadar basin include oil shale, carbonated rocks, evaporites, and tuff. These strata have been formed in their current position (authigenically) overgrown by massive layers of jadarite, which recognized boron-lithium silicate mineral of the zeolite family (Stanley et al., 2007). The white jadarite layers are reportedly several meters thick. This single jadarite deposition contains an estimated 3% of the world's lithium resource (Evans, 2012).

DID YOU KNOW?

Global annual production of lithium has been increasing since the mid-1990s; in 2020, it was about 82,000 metric tons (Mining.com, 2020).

REFERENCES

Barthelmy, D. (2014). Mineralogy database: Webmineral database. Accessed 10/31/22 @ https://webmineral.com.

Bradley, D., and McCauley, A. (2013). Mineral-deposit model for lithium-cesium-tantalum pegmatites. Accessed 12/12/22 @ https://pubs.er.usgs.gov/publication/ser20105070

Bradley, D., and McCauley, A. (2013). A preliminary deposit model for lithium-cesium-tantalum. (LCT) pegmatites. U.S. Geological Survey Open-File Report 2013-1008, 7 p.

Chao, J. (2021). Sizing up the challenges in extraction lithium from geothermal brine. Accessed 10/28/22 @ https://newscenter.ibl.gov/2021/11/25/sizingup...

Eggleston, T., and Hertel, M. (2008). King's Valley lithium project, Nevada, USA, NI 43-101 technical paper, prepared for Western Lithium Canada Corp., Project no 160237 [filing date January 21, 2009] Phoenix, Ariz., AMEC E & C Services Inc. December 15, variously paged. Accessed 11/5/15 @ http://www.sedar.com.

Evans, R.K. (2008). *An abundance of lithium.* Raleigh, North Carolina State University. Accessed 02/12 @ http://www.che.ncsu.edu/ILEET/phevs/lithium-availability/An_Abudnace_of_Lithium.pdf.

Evans, R.K. (2012). An overabundance of Lithium? Lithium Supply & Markets Conference, 4th, Buenos Aires, Argentina. January 23–25, 2012. Presentation, unpaginated.

Faure, G. (1998). *Principles and applications of geo-chemistry—A comprehensive textbook for geology students.* Upper Saddle River, NJ, Prentice Hall, 600 p.

Garrett, D.E. (2004). *Handbook of lithium and natural calcium chloride* (1st ed.). Boston, MA, Elsevier, 476 p.

Gruber, P.W., Medina, P.A., Keoleian, G.A., Kesler, S.E., Everson, M. P., and Wallington, T.J. (2011). Global lithium availability—A constraint for electric vehicles? *Journal of Industrial Ecology*, v. q5, no. 5, pp. 760–775.

Halpern, A. (2014).The lithium triangle. Accessed 10/31/22 @ https://web.archive.org/web/20180610055338.

Kunasz, I.A. (2006). Lithium resources, in Kogel, J.E., Trivedi, N.C., Barker, J.M., and Krukowski, S.T., eds., *Industrial mineral and rocks—Commodities, markets, and uses* (7th ed.). Littleton, CO, Society for Mining, Metallurgy and Exploitation, pp. 599–614.

Mining.com (2020). Visualizing the global demand for lithium. Accessed 11/01/22 @ https://www.mining.com/web/visualizing-the-global-demand-for-lithium.

Schwartz, M.G. (1992). Geochemical criteria for distinguishing magmatic and metasomatic albite-enrichment in granitoids—Examples from the Ta-Li granite Yichun (China) and the Sn-W Deposit, Tikus (Indonesia). *Mineralium Deposita*, v. 27, no. 2, pp. 101–108.

Simbol, Inc. (2013). Who we are: Pleasanton, Calif. Simbol, Inc. Accessed 05/09/14 @ http://www.simbolmaterials.com/who_we_are.htm.

Stanley, C.J., Jones, G.C., Rumsey, M.S., Blake C, Roberts, A.C., Stirling, J.A.R., Carpenter, C.J.C., Whitfield, P.S. Grice, J.D., and Lepage Y. (2007). Jadarite, $LiNaSiB_3O_7(OH)$, a new mineral species from the Jadar basin, Serbia. *European Journal of Mineralogy*, v. 19, no. 4, pp. 575–580.

Stillings, L.L., and Morissette, C.L. (2012). Lithium clays in sediments from close-basin, evaporative lakes in the Southwestern United States. *Abstracts with Program—Geological Society of America*, v. 44, no. 7, p. 210.

USGS (2017). *Lithium, professional paper 1802-K*. Reston, VA, United States Geological Survey.

Yaksic, A., and Tilton J.E. (2009). Using the cumulative availability curve to assess the threat of mineral depletion—The case of lithium. *Resources Policy*, v. 34, no. 4, pp. 185–194.

Part 2

Applications

Note: Lithium has uses in everyday life. There are several applications for usage of lithium at the present time. Many of these applications are briefly described in the following chapters. However, the final part is extensive and focuses on lithium that is used in the production of batteries.

DOI: 10.1201/9781003387879-5

4 Air Purification

INTRODUCTION

Lithium has a wide range of applications and can be used for a variety of purposes such as metallurgy, aircraft, air purification systems, nuclear weapons, pyrotechnics, treatment of mental disorders, manufacturing of glassware, optic, and several other uses or applications. Air purification uses of lithium are described in this chapter.

DID YOU KNOW?

Lithium with a specific gravity of 0.534 is about half as dense as water and the lights of all metals.

LITHIUM HYDROXIDE

Carbon dioxide is produced in our bodies whenever we consume food that is turned to carbon dioxide and released whenever we exhaled. Caron dioxide occupies only about 0.05% of the atmosphere. However, in a confined space such as inside a module for an LM-2500 gas turbine generator or prime mover when a fire or explosion occur, the model is flooded with carbon dioxide which is toxic and if anyone is stuck inside the module when the carbon dioxide dumps, the person will not last long because carbon dioxide is toxic. The carbon dioxide surrounds the victim inside the module and when the level of carbon dioxide concentration increases, the exposed victim will suffer certain symptoms:

- At 1%—will cause drowsiness
- At 3%—will cause stupor
- At 5%—will cause headache, dizziness, confusion
- At 8%—will lead to unconsciousness
- Above 8%—will cause death

Another example of carbon dioxide exposure that occurs within the confined space is in a spacecraft, space shuttle, and the space station. A good example of carbon dioxide buildup that must be removed occurs in the confined space of a spacecraft; the carbon dioxide must be removed from the spacecraft air by chemical processes. This is where lithium comes into play; actually, it is where lithium hydroxide is useful and necessary. Probably, the most well-known example of lithium hydroxide canister use occurred on the Apollo 13 mission.

What happened was that there was an explosion in the command module so the astronauts lived in the lunar module while the spacecraft returned to Earth.

DOI: 10.1201/9781003387879-6

The problem with moving to the lunar module is that it used round lithium hydroxide canisters while the command module used square ones (shows that inconsistency on relying on non-compatible devices instead using same or like devices is important—lesson learned). So, three astronauts were couped up in the lunar module breathing the air in a space designed for only two and they quickly used up the lunar module canisters of lithium hydroxide. The old saying is you can't put a square object into a round hole and the astronauts could not exchange the round canisters of lithium hydroxide for the square ones in the command module. So, the experts at Mission Control had to device a way to adapt the air flow of the lunar module though the square lithium hydroxide canisters. Mission Control experts (scientists, mostly) were able to rig a system using hoses, socks, plastic bags, and duct tape—this boot strap innovative way to remove carbon dioxide from the lunar module saved the astronauts from carbon dioxide death.

DID YOU KNOW?

Lithium has an average concentration of 20 ppm in the Earth's continental crust.

5 Electronic Devices

INTRODUCTION

Lithium is one of many elements and compounds used in several electronic devices. Figure 5.1 shows several elements or minerals or compounds that are used to manufacture an I-phone. Lithium is necessary for a number of things including for strategic, consumer, and commercial applications. The primary uses for lithium are in batteries (Note: An extensive discussion of lithium batteries is presented later in this text), ceramics glass, metallurgy, pharmaceuticals, and polymers.

The huge plus, advantage, and benefit of lithium use in electronic devices are its excellent electrical conductivity and very low density that allows it to float on water. These attributes make lithium an ideal component, for example, for battery manufacturing. Note that lithium is traded in three different forms: mineral compounds

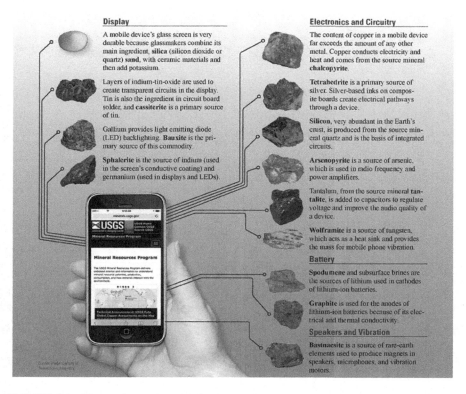

Display

A mobile device's glass screen is very durable because glassmakers combine its main ingredient, **silica** (silicon dioxide or quartz) **sand**, with ceramic materials and then add potassium.

Layers of indium-tin-oxide are used to create transparent circuits in the display. Tin is also the ingredient in circuit board solder, and **cassiterite** is a primary source of tin.

Gallium provides light emitting diode (LED) backlighting. **Bauxite** is the primary source of this commodity.

Sphalerite is the source of indium (used in the screen's conductive coating) and germanium (used in displays and LEDs).

Electronics and Circuitry

The content of copper in a mobile device far exceeds the amount of any other metal. Copper conducts electricity and heat and comes from the source mineral **chalcopyrite**.

Tetrahedrite is a primary source of silver. Silver-based inks on composite boards create electrical pathways through a device.

Silicon, very abundant in the Earth's crust, is produced from the source mineral quartz and is the basis of integrated circuits.

Arsenopyrite is a source of arsenic, which is used in radio frequency and power amplifiers.

Tantalum, from the source mineral **tantalite**, is added to capacitors to regulate voltage and improve the audio quality of a device.

Wolframite is a source of tungsten, which acts as a heat sink and provides the mass for mobile phone vibration.

Battery

Spodumene and subsurface brines are the sources of lithium used in cathodes of lithium-ion batteries.

Graphite is used for the anodes of lithium-ion batteries because of its electrical and thermal conductivity.

Speakers and Vibration

Bastnaesite is a source of rare-earth elements used to produce magnets in speakers, microphones, and vibration motors.

FIGURE 5.1 The various minerals/chemicals used for an I-Phone. Public Domain photo from USGS (2018).

DOI: 10.1201/9781003387879-7

from brines, mineral concentrates, and refined metal via electrolysis from lithium chloride. It is important to point out that lithium mineralogy is diverse; it is found a variety of pegmatite minerals such as spodumene, lepidolite, amblygonite, and in the clay mineral hectorite. At the present time (2022), current global production of lithium is dominated by pegmatite and closed-base brine deposits, but there are significant recourse in lithium-bearing clay minerals, oilfield brines, and geotherm brines (Bradley et al., 2017).

REFERENCES

Bradley, D.C., Stillings, L.L., Jaskula, B.W., Munk, LeeAnn, and McCauley, A.D. (2017). Lithium, chapter: K of Schulz, DeYoung, J.H., Jr., Seal, R.R., II, and Bradley, D.C., eds., *Critical mineral resources of the United States—Economic and environmental geology and prospects for future supply: U.S. Geological Survey Professional Paper 1802*, Washington, DC: Unites States Geological Survey, p. K1-21. https://doi.org/10.3133/pp1802K.
USGS (2018). Accessed 12/28/22 @ https://www/usgs.gov/data/lithium-deposits-united-states.

6 Optical Devices

INTRODUCTION

Lithium oxide, released from either lithium carbonate or spodumene, is used in the glass composition of optical products for spectroscopy. Lithium is used in spectacles especially made for infrared, ultraviolet, and vacuum ultraviolet ranges. Lithium is also used on container glass, flat glass, pharmaceutical glass, specialty glass, and glass products.

Why lithium?

Why lithium in spectacles? Lithium provides durability and is corrosion resistant or for use at high temperatures. Lithium increases the glass melt rate and lowers the viscosity and the melt process. Lithium process productivity for whatever product is enhanced with increased output and energy savings (USGS, 2014).

DID YOU KNOW?

- Lithium is more abundant than some of the better-known metals, including tin and silver.
- Lithium occurs in most rocks as a trace element,
- Worldwide consumption of lithium for use in spectacles, ceramics, and glass is 35%.
- Lithium is one of only three elements that were produced during Big Bang event.
- More than 300 million people benefit from a daily dosage of lithium carbonate.

LITHIUM FLUORIDE (LIF)

Lithium fluoride is manufactured for windows, prisms, specialized optics, and lenses in the vacuum special UV (ultraviolent), UV, visible, and infrared with a transmission range from 0.11 to 7 μm. Due to its refractive index, LiF can be found in electromagnetic radiation applications. Not only is lithium fluoride a benefit in optical industry, but it is occasionally used in focal lenses to focus higher energy proton beams. Lithium fluoride can also be found in mobile phones as a crystal oscillator (see Figure 6.1) which generates signals with faultless precision. Lithium fluoride is also used for X-ray monochromator plates.

DOI: 10.1201/9781003387879-8

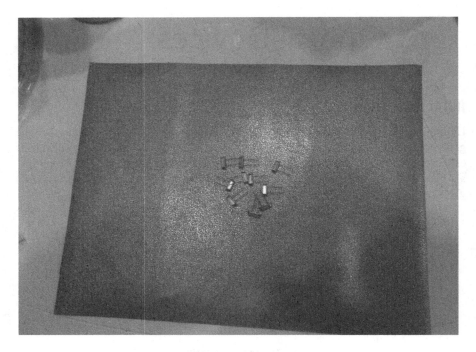

FIGURE 6.1 Type of crystal oscillator.

Okay, so what exactly is lithium fluoride?

Anytime if one needs to get the 411 on just about any chemical and chemical compound all one needs to do is check the safety data sheet (SDS) for the material—whatever the material might be. SDSs are a requirement of OSHA's 29 CFR 1910.1200 Hazard Communication Program; it is an informational document prepared by the manufacturer or importer of the hazardous substance.

So, according to a standard SDS, the following information is provided:

Lithium fluoride (SDS)	
Transmission range:	0.12–6 µm
Refractive index:	1.392 at 0.6 µm
Reflection loss:	5.2% at 0.6 µm
Absorption Coefficient:	5.9×10^{-3} cm^{-1} at 4.2 µm # 300K
Reststrahlen (residual radiation) peak:	25 µm
dn/dt:	-16×10^{-6} at 1.15 µm
Density:	2,639 g/cc

(Continued)

Lithium fluoride (SDS)	
Melting point:	848°C
Thermal conductivity:	11.3 W/K at 314 K
Thermal expansion:	37×10^{-6} K^{-1} at 283 K
Hardness:	Knoop 102 with 600 g indenter
Specific heat:	1,562 J Kg
Capacity:	
Dielectric constant:	0.1
Young's modulus:	64.97 GPa
Shear modulus:	55.14 GPa
Bulk modulus:	62.03 GPa
Elastic coefficient:	$C_{11} = 112$; $C_{12} = 46$; $C_{44} = 635$
Apparent elastic limit:	11.2 MPa (1,620 psi)
Poisson ratio:	0.33
Solubility:	0.27g/100g water at 20°C
Molecular weight:	25.94

DID YOU KNOW?

Modulus of Elasticity—the ratio of stress to strain, for stresses below the elastic limit. By checking the modulus of elasticity, the comparative stiffness of different materials can readily be ascertained. Rigidity and stiffness is very important for many machine and structural applications.

REFERENCE

USGS (2014). *Lithium—For harnessing renewable energy. Fact Sheet 2014-3035.* Washington, DC, United States Geological Survey.

7 Glassware

INTRODUCTION

This is the age of the microwave oven—when the meal is needed fast, microwave it. The glassware used to warm food in the microwave is made using lithium, which improves their thermal properties. The lithium used in ceramics and glass have played a major role in everyday life for decades. The thermal properties referred to include lithium's ability to reduce thermal expansion. Lithium used in glass-ceramics ensures that the molding is held together; that is, it is held in place and this prevents expansion because there is no thermal resistance.

The addition of lithium to ceramic glassware increases ceramic body strength. This leads to increased durability.

DID YOU KNOW?

The coefficient of thermal expansion for lithium is almost double in value than that of aluminum and four times that of iron due to which it is also used in glass and ceramics industries.

Lithium oxide (Li_2O) is used to reduce the melting point and viscosity of materials needed to manufacture high-strength glasses and ceramics.

In the fashioning of ceramic and other glass goods, artists and others have found that when they fashion their intended end product, whatever it might be, if they mold without lithium added, the base ceramic colors will influence their creation.

DID YOU KNOW?

Lithium was first discovered by Swedish chemist Johan August Arfwedson in 1817 when he analyzed mineral petalite ($LiAl(Si_2O_5)_2$).

Have you ever wondered why windows do not wilt under pressure? In practice, normal windows are double-glazed, with two glass panes separated by an air gap. The air gap is commonly filled with argon gas which provides excellent insulation and soundproofing properties. A thin layer of lithium added to the glass makes for a warm home in winter and a cooler house in the summer heat. Lithium ensures that the ceramics and glassware used today erode slowly because lithium reduces corrosion rates. It also works as a purifying agent which prevents oxides from forming on the surface.

DOI: 10.1201/9781003387879-9

8 Metallurgy

INTRODUCTION

In discussing lithium as used in metallurgy, we have to speak in two directions at the same time. The two directions are, number one, the metallurgy used in industrial use. In certain metallurgy processes, active lithium is used in a most important use as a scavenger element for the non-mental impurities—impurities such as oxygen, nitrogen hydrogen, carbon, and others. In another metallurgy process, lithium is used to improve melting behavior of aluminum oxides. Number two, metallurgy and other processes are used to extract lithium from resources. It is this process that we are interested in this book and as an example of the processes involved are detailed from the Federal Bureau of Mines investigation of two extraction lithium from the McDermitt Caldera clays.

McDERMITT CALDERA

McDermitt Caldera was introduced earlier in the book, but for clarity and understanding of following, the previous introduction is restated here.

The Kings Valley hectorite deposit, located within the McDermitt Caldera complex, and extinct volcanic center in northern Nevada are the locations where the most significant lithium-clay resources are found. During the Miocene the McDermitt area had extensive volcanism; at least five collapsed vents and resurgent domes have been recognized within the complex. The lithium clays occur in hydrothermally altered, volcanic-derived sediments of lakes that occupied the caldera (Glanzmand et al., 1978). A proven feasible method of lithium recovery that I have used in my research is by leaching the clay with sulfuric acid.

With regard to McDermitt Caldera's connection with lithium and metallurgy, high concentrations of lithium in fluviatile-lacustrine sediments constitute a promising resource. In size, McDermitt Caldera is roughly 45 km (28 miles; 147,638 ft) in diameter that is the result of not only volcanic activity but also subsidence and sedimentation, most of which occurred in the Miocene age. In the beginning, most of the sediments originally were vitroclastic (i.e., glassy rock fragments) and now consist chiefly of authigenic zeolites—means they were formed in their present position, clay minerals, feldspar, and quartz. Calcite occurs as thin beds, proturbulences and cement gypsum is present but scarce. The majority of the clay beds in the caldera contain well above the average lithium concentration. In the western and southern parts of the caldera, various smectite clay samples (i.e., they have large specific surface areas) have been found to contain up to as much as 0.65% Li and are associated with L-feldspar and analcime (a common feldspathoid mineral). In the northern section of the caldera, the clay beds are thinner and about 0.38% lithium (Ronov et al., 1991).

DOI: 10.1201/9781003387879-10

Okay, foundation, the basic geo-building blocks, and proper introduction to the McDermitt Caldera near the Nevada-Oregon border has been made. The question remains: Why the McDermitt Caldera?

Good question.

The Federal Bureau of Mines chose to study two different types of clay within the McDermitt Caldera. The study was all about finding the best methods of extraction of lithium from the two different lithium-containing clays. A summarizing study and a concise report of various lithium extraction processing characteristics of two lithium-containing clays from the McDermitt Caldera was reported by *911 Metallurgist* (2017).

While this text focuses on the real-world applications of lithium, it is also important to point out the work of Federal Bureau of Mines in studying the lithium extracting methods; it is detailed by and in *911 Metallurgist*'s *Extracting Lithium* (2017) accessed at https://.911/metallurgist.come/extracting-lithium—only a brief listing of the lithium extraction methods provided herein. For those wanting information of extraction techniques, you are referred to the website listed above.

DID YOU KNOW?

Other alkali metals are commonly found in plant material; however, lithium is found in spring and ocean waters.

LITHIUM EXTRACTION STUDIES

The Federal Bureau of Mines has determined that there are a number of approaches for extracting lithium from clays. Which of the extraction procedures to use in extracting lithium depends on the specific raw material being considered. Dozens of lithium extraction methods have been reported and many of them bench-tested, be advised that most of the processes have been established for pegmatite raw materials and may not be effective for extracting lithium from clay feed material. Earlier studies by the Bureau of Mines focused on lime-gypsum roast and chloride roasts for lithium extraction from spodumene and amblygonite. The techniques chosen for extracting lithium from clays were (1) water disaggregation; (2) hydrothermal treatment; (3) acid leaching; (4) acid baking-water leaching; (5) alkaline roasting-water leaching; (6) sulfate roasting-water leaching; (7) chloride roasting -water leaching; and (8) multiple-reagent roasting-water leaching.

REFERENCES

Glanzmand, R.K., McCarthy, J.H., and Rytuba, J.J. (1978). Lithium in the McDermitt caldera, Nevada and Oregon. Accessed 11/6/22 @ https://www.usgs.gov/publications/lithiyum-mcdermitt-caldera-Nevada-and-Oregon.

Ronov, A., Yaroshevuskly, A., and Migdisov, A. (1991). Chemical constitution of the earth's crust and geochemical balance of the major elements. *International Geology Review*, v. 33, p. 10.

9 Alloy

ALLOYS

Another common application of lithium is its use as an alloy with other elements to improve their mechanical and chemical properties. When lithium is alloyed with other elements like aluminum and magnesium, the strength of these elements is increased exponentially. Also, lithium reduces the weight of several alloyed elements, but again, at the same instant, increases their strength. Lithium's ability to increase strength and decrease weight easily explains why it is a common practice of alloying lithium with magnesium and aluminum. A couple of examples where lithium is alloyed with magnesium and aluminum are lithium-magnesium alloys used in making armor plating and lithium-aluminum alloys used in manufacturing the parts of airplanes and high-speed trains, and also in bicycle frames.

DID YOU KNOW?

Lithium is alloyed with copper and aluminum to save weight in airframe structural components.

DOI: 10.1201/9781003387879-11

10 Lubricating Grease

LITHIUM-BASED SOAP

Lithium is used in manufacturing lubricating greases. These greases are able to withstand high temperatures, minimalize friction, and are therefore used in various aircraft engines; its heaviest use as such occurred during World War II. These greases are prepared using lithium-based soaps (aka lithium stearate) because they have the ability to thicken the oils. In the process, the lithium-based soaps are made by heating the strong base lithium hydroxide (LiOH), with a fat, and the result is a product with a higher melting point in comparison with other alkali soaps.

LITHIUM STEARATE

As mentioned, lithium-based soaps are also known as lithium stearate. So, the question becomes: What is lithium stearate? Well, when we say "soaps," we are not talking about the type of soap most folks think about; instead we are talking about the metallic salts of fatty acids (also called soaps) that are used in various industries and for different reasons.

Soaps can have different properties based on the type of fatty acid and the length of the carbon chain and the alkali user. Fatty soaps with their longer chains are insoluble. To develop a hard soap, sodium hydroxide (aka caustic) is used as the alkali. Soft soap results when potassium hydroxide is used as the alkali.

It is interesting to note that different types of soap work for different purposes. For example, food-grade soap salts as food additives, while soap salts are also used as insecticides. These metallic salts of fatty acids are also applied as a stabilizer, depending on their properties, in the cosmetic and plastic industry. Because lithium is light, soft, and highly reactive it can be used in various alloys to form both inorganic and organic compounds, such as lithium hydride, lithium carbonate, lithium hydroxide, and also one of those compounds is lithium stearate.

So, the question is: What is Lithium Stearate?

Lithium stearate is a salt of stearic acid, which is derived from lithium hydroxide with cooking tallow or other animal fat. In practice, lithium stearate is used as a general-purpose lubricating grease that provides high resistance to water. Because of its properties, lithium stearate has applications as a corrosion inhibitor in petroleum. The automotive, heavy machinery, and aerospace industries use lithium stearate because it can be applied both in high and low temperature. Note that because it can be used as a general-purpose lubricating grease, it is commonly used in the cosmetic industry as a stabilizer and also in the cosmetic industry.

So, the next question is: Where is lithium stearate used, applied?

Lithium stearate is used in many different applications from acting as a great sealant to use as an anti-rust and corrosion agent. It also has excellent high-temperature applications. Natural and synthetic oils use lithium stearate as a thickening agent.

DOI: 10.1201/9781003387879-12

Lithium is also used in organic synthesis at both laboratory and industrial levels. Another application of lithium chloride and lithium bromide is their use as sorbents for gas streams—this beneficial application is possible because of the hygroscopic nature of lithium chloride and lithium bromide. Lithium hydroxide and peroxide are used in submarines, spaceships, and other closed spaces to remove carbon dioxide, purifying the air.

DID YOU KNOW?

Lithium salt-based lubricant was patented in 1942 in the United States by Clarence E. Earle.

LITHIUM HYDROXIDE

Lithium is a critical component of the lithium-ion batteries (discussed in detail in Part 3 of this book). At the present time, most producers use lithium carbonate for battery applications. However, as the technology advances, it is found that a sea change is occurring in the transfer of usage from lithium carbonate to lithium hydroxide (LiOH). The compound, lithium hydroxide, is obtained from the reaction of lithium carbonate with calcium hydroxide. It is generally used for making soaps (aka lithium salts) of stearic and other fatty acids.

So, after introducing lithium hydroxide, the question becomes: What is it used for?

Well, as previously mentioned, lithium hydroxide is used in the manufacture of lithium salts (soaps) of steric and additional fatty acids, these soaps are widely used as thickeners in lubricating grease. This lithium grease has several properties including high temperature and water resistance and it can also sustain extreme pressures, making it suitable for various industrial applications. Its primary application is in the automotive and automobile industry. Also, lithium products are used as additives in the electrolyte of alkaline batteries.

The bottom line: With the ever-increasing manufacture and sales of electric vehicles (EVs) lithium compounds are expected to increase exponentially in their use.

DID YOU KNOW?

Lithium has melting point of 180.5°C, the lowest among all metals but the highest compared to other alkali metals, and is highly reactive and possesses high flammability tendency similar to other alkali metals.

Note: Only a few of the lithium applications have been discussed to this point. Other applications include military and aerospace use, welding and other process applications such as coolant in nuclear processes, and also as a fuel in rocket propulsion. At this point in the book, the next two major applications, medical and battery power, because their use is so widespread and spreading, are the focus of the book.

Part 3

Medicinal Use of Lithium

11 Mania Treatment

INTRODUCTION

For more than 70 years, lithium has been used clinically for treatment of bipolar and other psychiatric disorders, such as schizoaffective and bipolar illness. When lithium is augmented with antidepressants, it is useful in treatment-resistant unipolar depression. Lithium is also used effectively in treating cluster headaches and in improving chemotherapy-induced neutropenia (i.e., low white blood cell lever). Other uses of lithium in psychiatry and medicine have been reported and reviewed. However, the use of lithium may be limited by acute and chronic toxic side effects. The side effect has to do with central nervous system (CNS) dysfunction—the degree of toxicity parallels the extent of CNS dysfunction. Experience has shown that chronic toxic manifestations effect cardiac, renal, and endocrine systems. In fetu exposure may cause a rare developmental abnormality; it may be teratogenic (el-Mallakn, 1990).

HISTORICAL PERSPECTIVE

In 1949, an Australian psychiatrist, John Cade (1929–1996), unexpectantly initiated a new era in psychiatric treatment by using lithium carbonate to treat mania. Cade's use of lithium sprung from the hypothesis that major mental illnesses might be associated with deficiencies or excesses of unidentified chemical substances, including accumulations of nitrogenous metabolites. Cade gave lithium carbonate to laboratory animals to limit toxicity of test substances including uric acid and noted calming and other behavioral changes. Afterward, Cade reported on the beneficial effects of treating ten patients with lithium carbonate (an unproved but medically accepted treatment for gout) for mania and on risk of discontinuing such treatment (Cade, 1949, 1999). These promising initial results are now widely considered a groundbreaking discovery, this is the case even though this novel, actual, and effective treatment was not immediately adopted by psychiatry. Moreover, it is interesting to note what Cade himself perceived, "a discovery by an unknown psychiatrist without research training, working in a small hospital for the chronically mentally ill, with techniques and negligible equipment, could not attract much attention" (Cade, 1999). Also of note, unfortunately, was a failed experiment with lithium chloride as a substitute for table salt in patients with congestive heart failure. Moreover, it was reported in 1949 that several cases of severe, acute intoxication occurred connected with use of lithium salts again as it was used as a substitute for sodium chloride (table salt). Basically, it was determined that experience was required to learn how to use lithium safely. And to accomplish this, concentration level in blood needs to be measured (Amdisen, 1967; Baldessarini, 2013; Bauer and Gitlin, 2016). Cade's discovery was significant not just because it added an important new agent to the psychopharmacologic armamentarium but because it illustrated

DOI: 10.1201/9781003387879-14

the triumph of the scientific method; this was important because it was a time when psychiatry was in danger of losing sight of science.

DID YOU KNOW?

Dr. John Cade fought for Australia during World War II. Although trained as a psychiatrist, he served as a military surgeon. He served in Singapore, and after its fall, he became a prisoner of war from September 1941 to September 1945. While he was imprisoned, he observed various forms of behavior by his fellow prisoners. Of the various forms of behavior Dr. Cade observed were some fellow inmates who not only acted strange but also exhibited vacillating behavior (i.e., some inmates wavered in mind and/or opinion). Dr. Cade surmised that is was a toxin of some type that was affecting their brains and actions and when it was eliminated through their urine, they lost their symptoms.

After the war, Dr. Cade took up a position at a hospital in Melbourne. In an unused kitchen, Dr. Cade conducted crude experiments which eventually led to the discovery of lithium as a treatment mania (aka bipolar disorder). After ingesting lithium himself to ensure its safety in humans (Cade, 1999), Dr. Cade conducted a trial of lithium citrate and/or lithium carbonate on some of his patients diagnosed with mania, melancholia, dementia, with exceptional results. Because his observations made during his treatment of his patients were so strong, he speculated that the cause of mania is the result of lithium deficiency (Cade, 1949).

After Dr. Cade's findings about the use of lithium were published, Professor Erick Stromgren (1909–1993), a well-known academic psychiatrist at a university medical center in Denmark, read about Cade's work with lithium for mania (today called bipolar) and asked one of his colleagues to replicate the Australian findings. Cain's colleague Morgen Schou (1918–2005) basically took the ball, so to speak and ran with it in his further review of Cade's work, and after extensive study and experimentation, he was able to establish safe work practice in handling lithium and the effective clinical use for the treatment of manic-depressive illness.

During the time when Morgen Schou was conducting his research about the use of lithium, he collaborated with Poul Baastrup (1818–2001), another Danish psychiatrist, and they established the most important effect of lithium—the ability to prevent recurrence of maniac-depressive illness (Baastrup, 1964; Baastrup and Schou, 1968). Anyway, in Zurich, Jules Angst and his collaborators confirmed the prophylactic (aka preventive) benefits of lithium treatment (Angst et al., 1970). What was needed after determining the benefits of lithium treatment was, using the precepts of the scientific method, to come up with proposed methods of clinical-trial design suitable for testing long-term prophylactic effects of a treatment such as lithium. The proposed methods were studied and put forth by Paul Grof, Angst, and other colleagues (1970).

With the passage of time, that included continued study and usage of lithium became accepted in clinical practice around the globe, although it is important to point out that there were skeptics—and this strong skepticism is good in science and medicine because actual verification of the efficacy and any new scientific theory or practice must be able to survive skeptics and regulation replication of the original findings; otherwise, a theory or finding will not hold water, so to speak, because it can't be repeated, consistently. Most of the early skepticism was based on reasonable considerations such as about lithium treatment having long-term efficacy (Blackwell and Shepherd, 1968). The reasonable considerations included the requirement for evidence from blinded, randomized, controlled trials (the Scientific Method at work).

Okay, if you are not familiar with the scientific method, consider that in *science*, the scientist uses the scientific method, grounded in experimentation. Scientists conduct controlled experimentation, which tend to be reductive (they isolate the problem to a single variable, sometimes missing the big picture). Experimentation takes time. Rushing it can—and often will—invalidate the results. Effective scientists must remain objective, value-free, bias-free. *Scientists* use problem-solving techniques. They start with a human-caused problem and consider the human values pertinent to the problem. The environmentalist and health researcher consider human values, which are not value-free, not objective, and not bias-free (neither is society).

Scientists define natural system structure, function, and behavior that may or may not have direct application to a particular environmental problem. *Scientists* define a process for solving environmental problems.

Scientists propose hypotheses based on past observations and use the scientific method to continue questioning and testing to establish the validity of hypotheses. Environmental scientists use problem-solving to propose future-directed solutions and continually evaluate and monitor situations to improve the solutions.

Scientists are interested in knowledge for their own sake; or in applied sciences, applications of knowledge found through a precise and thorough process that may or may not solve environmental problems. Scientists are interested in finding the best solution (sometimes before all the facts are in) to "actual" environmental problems within a particular social setting.

Okay, back to the evolution of lithium and its efficacy in the treatment of medical issues. The precepts of the scientific method in the first trial in obtaining evidence from blinded, randomized, and controlled trials, which worked to convince the scientific and clinical community (Baastrup et al., 1970). And finally, lithium treatment received regulatory approval by the U.S. Food and Drug Administration (FDA) in 1970 for treatment of acute mania, and in 1974, as the first—and for several years, the only—approved treatment for prevention of recurrences in bipolar disorder. The seemingly slow acceptance of lithium for medical purposes can be blamed on two problems. First, lithium has potentially toxic effects. And second, lithium is a naturel product meaning that it can't be patented; thus, the commercial interest had limited support for research or clinical promotion by pharmaceutical interests. Simply, without the ability to patent lithium meant that no one company or corporation could harness the right to charge whatever they wanted for the product.

So, some means had to be found to make lithium's clinical applications feasible. Again, with passage of time, this was accomplished by the introduction of sensitive, reliable, quantitative methods of monitoring lithium concentrations in serum initially with flame spectrophotometry and later with atomic absorption, electrochemical, and other detection methods (Tondo et al., 2019).

It is important to note that established clinical use of lithium needs to be monitored. It's all about dosing; the safety and health of those receiving lithium doses. Thus, monitoring is essential and required because of the narrow margin between the safe and potentially toxic dose of lithium, or its therapeutic index (TI) which is a ratio that compares the blood concentration at which a drug becomes toxic and the concentration at which the drug is effective. With regard to lithium toxicity, it is closely related to serum lithium concentrations and can occur at dose close to therapeutic concentrations. Undeniably, lithium remains unique in not being dosed adequately by the mg dose of drug given per day, but instead by achieving serum concentrations 10–14 hours after the last taken dose (this is when patient is in the most stable range of the day) in the range of 0.5–1.0 mEq/L, while being aware that earlier, daily peak concentrations can be two to three times higher (Amdisen, 1967; Baldessarini, 2013; Bauer and Gitlin, 2016; Perugi et al., 2019).

DID YOU KNOW?

Each 5 mL of Lithium Oral Solution contains 8 mEq of lithium ion (Li$^+$) which is equivalent to the amount of lithium in 300 mg of lithium carbonate.

Table 11.1 presents an early timeline in the history of lithium from discovery to USEDA approval.

TABLE 11.1
Timeline in the medical history of lithium

Year	Event
1817	Johan August Arfwedson discovers lithium
1843	Alexander Ure introduces lithium in modern medicine
1855	William Thomas Brande fully isolates lithium
1879s	William Hammond—anecdotal evidence of lithium bromide in treatment of acute mania
1890s	Carl Lange—systematic use of lithium in the acute and prophylactic treatment of depression
1900s	Toxicity reports—weakness, tremor, diarrhea, vomiting and deaths
1932	Lithium disappears from British Pharmacopoeia
1940s	Use of sodium substitute in low-sodium diets
1949	Removal from American markets following reports of severe intoxication
1949	John F.J. Cade—use of lithium in acute mania
1850–1974	Intense clinical research into safety and efficacy of lithium
1968	American Journal of Psychiatry recognizes the clinical significance of lithium
1979	USFDA approval for treatment of mania
1974	USFDA approval for maintenance therapy of patients with mania

Source: Adaptation from Johnson and Gershon (1999).

BIPOLAR DISORDER[1]

Before presenting a discussion of lithium for clinical use, it is important to discuss bipolar disorder that the medical lithium is designed to treat. One of the best descriptions of bipolar disorder is provided by the American Psychological Association and is adapted from *Encyclopedia of Psychology*:

> Bipolar disorder is a serious mental illness in which common emotions become intensely and often predictably magnified. Individuals with bipolar disorder can quick swing from extremes of happiness, energy, and clarity to sadness, fatigue, and confusion. These shifts can be so devastating that individuals may consider suicide. All people with bipolar disorder have manic episodes—abnormally elevated or irritable moods that last at least a week and impair functioning. But not all become depressed.

THE GOLD STANDARD

Today, lithium treatment remains the "gold standard" of treatment for preventing recurrences in bipolar disorder, both types 1 and II. According to the American Psychiatric Association (2022), there are five different types of bipolar disorder. Bipolar disorder actually consists of five different types:

- Bipolar I disorder—with mania and major depression
- Bipolar II disorder—with depression and hypomania
- Cyclothymic disorder—mood disorder consisting of numerous alternating periods of hypomanic and depressive symptoms
- Other specified bipolar and related disorder—is diagnosed when you have the symptoms of bipolar disorder, but they don't fit into other bipolar categories.
- Unspecified bipolar and related disorder—used when doctors do not have enough information.

Earlier it was mentioned that lithium treatment works to prevent patients with bipolar or major depressive disorder from exhibiting suicidal behavior (***note***: lithium and suicide is discussed in further detail later). However, it has slowly become less widely utilized, predominantly for mania, largely due to more robustly promoted and more rapidly effective alternatives which do not require blood tests (Tondo et al., 2019). There are lithium being used at the present time and these include drugs developed for other purposes, include certain antiepilepsy agents such as carbamazepine, lamotrigine, sodium valproate, and most antipsychotics. These alternatives do not come without adverse effects including metabolic syndrome (weight-gain with diabetes, high blood pressure, and increased lipids in the blood), insulin resistance, abnormal movements, and cognitive dulling with antipsychotics, as well as markedly increased risks of birth defects include spina bifida and several cardiac anomalies during pregnancy in association with valproate and carbamazepine (Patel et al., 2018). It is important to point out that concerns about the safety of lithium have not entirely disappeared, despite long-established standards for its sage use with monitoring of its serum concentrations.

Currently, fears about the safety of lithium treatment are most often direct to putative renal toxicity (i.e., the sudden malfunction of the kidneys) with its long-term use. The truth be told, such effects are infrequent and usually can be expected by increasing serum concentrations of creatinine or declining creatinine clearance (McKnight et al., 2012; Baldessarini, 2013; Bauer and Gitlin, 2016; Haussmann et al., 2017; Perugi et al., 2019). Neilson et al. (2018) point out that this adverse effect of lithium is modest and greatly overlaps with age-associated declining renal function. A sign of declining renal function (at least one elevated serum concentration of creatinine) was recorded in about 30% of subjects treated with lithium for 15 years or more and aged 55 years or older (Tondo et al., 2017).

Some patients and families may feel the lithium treatment has a "stigmatizing" or "branding" effect, as lithium is identified as a medication for severely ill patients, in contrast to seemingly less stigmatizing anticonvulsant, antidepressant, antipsychotic, and other psychotropic medicines (Baldessarini, 2013; Bauer and Gitlin, 2016). In some countries, particularly the United States, strong promotion of substitute and patented treatments, sometimes more promptly acting against mania, has led to significant displacement of lithium for acute mania. This is not to say that this trend has issued and meant the death nil toward using lithium; it continues to hold a major position among treatments for bipolar disorder internationally, especially for long-term prophylactic treatment, and it tends to be used for longer times than most alternatives (Baldessarini et al., 2007, 2008, 2019).

The bottom line: lithium still remains the "gold standard" when it comes to treating bipolar disorders.

DID YOU KNOW?

Intermittent episodes of manic and depression are the characteristics of bipolar disorder; without treatment, 15% of patients commit suicide (Sharma and Markar, 1994). So, it has been ranked by the World Health Organization as a top disorder of morbidity and lost productivity (Dusetzina et al., 2012).

EFFICACY AND EFFECTIVENESS OF LITHIUM TREATMENT

When some people hear about someone or anyone who is receiving lithium treatments, they may instantly jump to conclusions that the patient is receiving the lithium because she or he is suffering from being overexcited, hyper, agitated, hectic, frenzied, hysterical, frantic, weird, or possessed—manic. And excluding the weird and possessed descriptors, the other ailments listed are certainly "episodes" experienced by some patients and many of these have been treated with lithium. Although lithium is effective in treating these listed episodes, its primary value is as a "mood-stabilizing" agent.

Mood-stabilizing agent?

Yes.

So, before going forward with this discussion, let's first discuss mood episodes. People with bipolar disorder may experience periods of unusually intense emotion, excitement and/or sensation, changes in vigor/energy and action, and uncharacteristic (abnormal/strange) behaviors. In order to fully understand how mood disorders are defined, it is first important to understand the concept of mood episodes. According to the American Psychiatric Association and its published *Diagnostic and Statistical Manual of Mental Disorders* (2022), there are four kinds of mood episodes: major depressive, manic, hypomanic, and mixed. A major depressive episode is a period of at least two weeks during which you experience five or more depressive symptoms nearly every day and they impact the person's functioning. Manic episodes are characterized by a persistently elevated, expansive, or irritable mood—sometimes unusually angry and lasts at least one week to be diagnosed. Hypomanic episodes are less severe manic episodes and are called hypomania. Hypomanic episodes only need the person's mood disturbance to occur throughout at least 4 days. Mixed episodes are essentially a combination of manic and depressive episode that become superimposed so that the symptoms of both are present, at different times, during the day (Mentalhelp, 2022).

Okay, let's get back to mood-stabilizing agent which is simply an aim at long-term prevention of recurrences of acute illness episodes in bipolar patients (Baldessarini, 2013; Bauer and Gitlin 2016; Haussmann et al., 2017; Perugi et al., 2019), with greater assistance against repetition of mania than for bipolar depression (Kleindienst and Greil, 2000), as the case also with different treatments (Baldessarini, 2013; Forte et al., 2015). Grof (2006) found satisfactory mood stabilization can be attained over 6–12 months by about two-thirds of lithium-treated bipolar disorder patients, and an excellent response was found in one-third of such patients. Fairly recent wide-ranging reviews considered the value and benefits of various proposed mood-stabilizing treatments, including lithium, against recurrences of mania and depression in bipolar disorder patients, in more than a dozen of placebo-controlled, randomized, long-term trials carried out for an average of 1.5 years (Geddes et al., 2004; Popovic et al., 2012; Vieta et al., 2011; Miura et al., 2014). Where these lithium versus a placebo studies were conducted, it was found that long-term risk of new manic or depressive episodes was lower with lithium than with placebo. The only fly in the ointment in these studies was the finding that the benefit the lithium versus placebo was that lithium treatment was greater against new episodes of mania than of depression. But lithium is not alone in this instance because anticonvulsants and antipsychotics are also notably limited in their effectiveness against recurrences, the only exception was lamotrigine, it was effective for long-term treatment (Geddes et al., 2004, 2010; Poon et al., 2021; Baldessarini, 2013; Vasquez et al., 2013; Forte et al., 2015). Then again, in the 1970s and more recently, some randomized, double-blind trials found that long-term treatment with lithium also may reduce recurrences in unipolar major depressive disorder (Abou-Saleh et al., 2017; Tiihonen et al., 2017; Undurraga et al., 2019). Also, lithium appears to have value in supplementing antidepressant treatment, especially during the episodes of unipolar major depression that respond unsatisfactorily to antidepressant treatment. The majority of studies supporting this application have involved older, tricyclic (TCA) antidepressants, but similar effects may occur with modern antidepressants as well (Austin et al., 1991; Bauer and Dopfmer, 1999; Bauer et al., 2003; Alevizos et al., 2012; Bauer and Gitlin, 2016. Undurrga et al., 2019). Finally, it has been suggested that lithium occurring naturally in drinking water may lower the incidence of dementia, but his finding needs to be confirmed (Kessing et al., 2017).

ALTERNATIVE RESPONSE TO LITHIUM

Mood-stabilizing and prophylactic benefits of lithium may be particularly evident in patients with typical bipolar disorder. Note that typical bipolar disorder includes an episodic clinical course before treatment, a family history of the disorder and favorable response by a family member, lack of other co-occurring psychiatric illnesses, and an illness coarse sequence characterized by manic or hypomania followed by depression and then a stable or euthymic interval (i.e., euthymic is the state of being in neutral mood, euthymia; aka MDI pattern; opposite DMI—depression-mania-euthymic-interval) (Koukopoulos et al., 2013; Malhi et al., 2017; Yatham et al., 2018). In particular, patients with an MDI course type, as is likely in type I bipolar disorder, triggered by overuse of antidepressants have shown a 29% (C1 8–40%) better response to lithium than those with a DMI course, such as those in type II bipolar disorder, often triggered by overuse of antidepressants (Kukopulos et al., 1980, 2013). Note that the association with the type of illness course is consistent with the view that depression-prone bipolar disorder patients, generally, respond less favorably to treatment than mania-prone patients (Vieta et al., 2009; Baldessarini et al., 2012). Note that possibly from the first-lifetime episode or cycle both, the DMI versus MDI recurrence pattern, as well as the tendency to a long-term excess of mania over bipolar depression, may be identifiable very early on the illness history, (Baldessarini et al., 2012; Koukopoulos et al., 2013; Forte et al., 2015).

Among patients with relatively complicated forms of bipolar disorder, such as with rapid cycling, psychotic features, co-occurring anxiety syndromes, or substance abuse, as well as in depression-prone cases or those following the DMI course pattern, lithium is less effective in preventing recurrences of bipolar depression than mania. Notably, the same limitations hold true for alternative treatments, including anticonvulsants (Baldessarini et al., 2002; Tondo et al., 2003; Bauer and Gitlin, 2016), and no alternative treatment appears consistently to outperform lithium for long-term maintenance treatment (Geddes et al., 2004, 2010; Baldessarini, 2013; Pacchiarotti et al., 2013; Vasquez et al., 2013; Bauer and Gitlin, 2016; Malhi et al., 2016; Yatham et al., 2018). Also, as for all maintenance treatments, an encouraging clinical response is associated with faithful adherence to the treatment regardless of current absence of symptoms, absence of early or current adverse life events, adult age at onset, good social support, and absence of substance abuse or other co-occurring psychiatric disorders including personality disorders (Yatham et al., 2018).

ON AND OFF MEDICAL TREATMENT AND MONITORING

There is a clear lack of consensus as to whether indefinitely continued maintenance treatment should be started routinely after an initial manic episode. A moderate view would initiate long-term treatment after a second episode of mania since the first might be followed by another only after several years or may be of only moderate severity and duration. After an acute episode of mania or bipolar disorder, patients are typically continued on lithium or an alternative treatment for at least some time (usually months) following recovery, with reassessment of the need to continue thereafter. Right after a first manic episode presented with severe symptoms, that required hospitalization, or involved suicidal risk or prolonged duration lithium treatment probably should be started. Generally speaking, the practice of early long-term intervention and follow-up are encouraged after a manic episode, and especially of juvenile onset, because of the adverse impact of bipolar illness on patient's educational, occupational, and social functioning (Kessing et al., 2017).

For the reason of frequent lack of timely recognition and diagnosis of bipolar disorder, especially of type II, initiation of long-term treatment with lithium or other mood-stabilizing treatments typically does not occur for 5–10 years from illness onset, and even longer following juvenile onset (Post et al., 2010; Kessing et al., 2017). Note that it is such delay, with associated morbidity, that in turn leads to disability and risk of suicide. Even so, some studies have found that neither such delays nor the number of recurrences before treatment had a measurable impact on the likelihood of responding once mood-stabilizing treatment was initiated (Baethge et al., 2003; Bratti et al., 2003; Berghofer et al., 2008).

Substantial risk of illness recurrences, need for hospitalization, and of suicidal behavior result when long-term lithium treatment is discontinued, especially if the discontinuation is rapid or abrupt. The risk is particularly high in the 6–12 months after discontinuation (Faedda et al., 1993; Baldessarini and Tondo, 2018).Worth mentioning, recurrences of bipolar disorder following discontinuation of maintenance treatment may be more severe and occur much sooner than lithium is discontinued rapidly (<15 days) or abruptly (Faedda et al., 1993; Baldessarini, 2013). The bottom line is that this effect not only reflects lack of treatment but may be a response to treatment discontinuation itself as a major stressor (Baldessarini and Tondo, 2018).

STANDARDIZED MORTALITY RATIO

The standardized mortality ratio (SMR) for all-cause mortality as compared to the general population has increased for patients with bipolar disorder—this also includes higher rates of suicide at young ages and of fatal medical illnesses in late year (Osby et al., 2001; Staudt-Hansen et al., 2019). Note that the SMR is a quantity, the ratio of observed deaths in a study group to expected deaths in the general population (multiplying the ratio by 100 will produce a percentage). SMR for suicide as the cause of death is far higher with bipolar disorder and severe, recurrent, non-polar major depression (chiefly involving psychiatric hospitalization) than other psychiatric disorders, up to 20 times higher than the international general population rate of approximately 15/100,000/year (0.015%/year) (Harris and Barraclough, 1997; Osby et al., 2001; Baldessarini, 2013). What this boils down to is that the rate of suicide attempts is at least 20–30 times greater than threats of suicide in the general population but the rate of suicides/attempts is much lower among patient with a more significant mood disorder. The significance of this ratio is that it indicates greater lethality of means and intent in major mood disorder patient with a similar risk of suicidal behavior in types I and II bipolar disorder (Tondo and Baldessarini, 2016).

Original considerations that lithium might contribute to suicide prevention date to the early 1970s, followed by important contributions supporting this hypothesis over the next two decades (Coppen et al., 1983). A study of suicidal risk in 68 patients with various major affective disorder diagnoses and at least one suicide attempt found a rate of suicides or attempts during 8.0 years of lithium treatment of 1.1%/year, with a highly significant increase to .0%/year following discontinuation of lithium treatment (Muller-Oerlinghausen et al., 1992). Note that in 360 types I or II bipolar disorder patients before, during, and following discontinuation of long-term lithium monotherapy, rates of suicide and life-threatening attempts were 6.4 times lower during lithium treatment that either before (only attempts) or long after treatment. Within several months after discontinuing lithium maintenance treatment, the risk of suicidal acts increased 20-fold, but afterward it fell back to the same level encountered before lithium treatment had started (Tondo et al., 1998). Also, early suicidal risk following discontinuation of long-term treatment with lithium was twice higher following abrupt or rapid versus more gradual discontinuation of lithium over at least 2 weeks.

In an orderly appraisal of 12 studies, the pooled rate of suicides and attempts was 8.9 times lower with lithium treatment ($p < .00001$) (Tondo et al., 2001). Moreover, Guzzetta et al. (2007) point out that a review of 8 studies on unipolar recurrent major depression patients found that long-term lithium treatment was again associated with a substantial reduction of risk of suicides and attempts (by better than 70%) among patients treated with lithium compared to other alternatives, mainly antidepressants and anticonvulsants. Note that similar effects on suicidal behavior have not been achieved with several proved or putative (i.e., supposed) mood-stabilizing anticonvulsants (including carbamazepine, lamotrigine, and valproate) (Thies-Flechtner et al., 1996; Goodwin et al., 2003; Baldessarini and Tondo, 2009; Smith et al., 2009). Also note that antipsychotic drugs have not been tested for such effects in bipolar disorder patients. Several studies and quantitative review have supported an association of reduced suicidal risk during long-term treatment with lithium in bipolar disorder patients (Angst et al., 2005; Cipriani et al., 2005, 2013; Kessing et al., 2005; Muller-Oerlinghausen et al., 2006; Lauterbach et al., 2008). Possibly an antisuicidal effect of lithium treatment is at least partly independent form its mood-stabilizing effect (Aherns and Muller-Oerlinghausen, 2001; Manchia et al., 2013). If this proves to be the case, patients achieving only a partial clinical response to lithium treatment might still benefit from its antisuicidal effects (Aherns and Muller-Oerlinghausen, 2001).

It is safe to say that overall the findings summarized here appear to provide strong and quite consistent support for the hypothesis that lithium treatment may have a special role in reducing suicidal risks. While it is true that these findings are encouraging and unusual, it must be emphasized that such an effect of lithium treatment has not yet been definitely proved in prospective, randomized trials in which suicidal behavior is an explicit outcome. The bottom line: largely for that reason, lithium has not been accepted by the FDA as a specific indication to prevent suicidal ideation or behavior.

MANAGING LITHIUM TREATMENT

Note that the following guidelines for managing lithium treatment are based on several recent publications and the extensive clinical experience (Baldessarini, 2013; Bauer and Gitlin, 2016; Haussmann et al., 2017; Malhi et al., 2017; Perugi et al., 2019). As mentioned, lithium is used mostly as a long-term treatment to prevent mood disorder recurrences. With regard to acute mania, lithium treatment has been largely displaced in favor of some anticonvulsants and modern antipsychotic drugs, which act more rapidly and whose target doses can be reached within a few days. Particularly, the most common current treatment for a manic episode is with modern antipsychotic agents for several months, with lithium introduced adjunctively or continued long term by itself as a preventive treatment (Yatham et al., 2018).

According to Malhi et al. (2017), lithium should be taken regularly as prescribed. Moreover, a daily single dose after the evening meal is convenient, preferable with slow-release formulations in relatively young, otherwise healthy patients. Now, for older or weakened patients, and users of high daily doses (over 1,200 mg of lithium carbonate [32 mEq—milliequivalent]), it may be safer to take divided daily doses. Whenever there is a need to change brands or salt form of lithium needs changing, this should be done gradually discontinuing the first preparation as the second is introduced and gradually increased. The most favorable amount of lithium to be taken is centered on clinical response and measured blood levels of lithium which guide the dose of lithium. Typically, it is 1 week after the start of lithium treatment that blood assays of lithium are obtained, then monthly in the first 3 months. Afterward, when the patient is considered stable, blood tests may be done every 6 months depending on age general health, 10–14 (optimally, 12) hours. If a dose is switched, 5–7 days should pass prior to measuring the blood level to allow tissue dispersal to become stable. It is the clinician who decides optimal doses of lithium and is dependent on the patient's age, general health, type of bipolar disorder, symptom severity, and frequency of recurrences. Optimal daily trough-blood concentrations (C_{trough}) of lithium for long-term treatment is typically between 0.50 and 0.60 and 0.80–1.00 mEq/L. Note that some patients may require higher concentrations, whereas for others, lower concentrations may suffice and be better tolerated. Higher serum levels of lithium may be required, for example, for young patients with severe manic or psychotic symptoms (delusions and hallucinations) or patients with short intervals between episodes, whereas lower concentrations are often used and better tolerated by elderly patients.

DID YOU KNOW?

A trough level or trough concentration, in medicine and pharmacology, is the concentration reached by a drug immediately before the next dose is administered (AGAH, 2004).

In patients with fever above 38°C (100.4°F), dehydration or diarrhea the lithium dose should be decreased by half or held for some days. Also, extra caution is advised in taking lithium during treatment with certain medicines that can slow its renal elimination and increase risk of toxic effects, and if a low-sodium diet is required for medical reasons. These drugs include the commonly used non-steroidal and inflammatory drugs (NSAIDs) such as ibuprofen or nimesulide if they need to be taken more than occasionally. If such drugs are taken regular for extended periods of time, lithium blood measurements are recommended. Note that acetaminophen/ paracetamol is a safer treatment for pain or fever than an NSAID. Angiotensin-converting enzyme (ACE) inhibitors (anti-hypertensives) and other drugs used to treat cardiac arrhythmias, and most diuretics should be used with caution. When using lithium with any of these medicines, serum lithium levels may rise and should be monitored frequently. Moreover, lithium treatment should be stopped 48–72 hours before surgery that requires general anesthesia, and while a patient does not drink of eat as usual. No interruption is needed for surgeries requiring local anesthesia. When the patient is well hydrated, lithium treatment may be resumed (Huyse et al., 2006). As a final point lithium should be stopped during electroconvulsive treatment (ECT) as a precaution to prevent possible neurological symptoms (delirium) (Tsujii et al., 2019).

HARMFUL AND ABNORMAL RESULTS

When a patient is treated with lithium, there often are reactions that are unfavorable. These can include tremor (a dose-dependent effects, which can be treated with low doses of the centrally active, beta-adrenergic blocker propranolol, high doses of vitamin B6, or with dose reduction if possible), nausea, fatigue, increased appetite, increased white blood-cell count, thirst, and increased frequency of urination (polyuria). This particular symptom may respond to cautiously added, small doses of the diuretic hydrochlorothiazide (which can also increase serum concentrations of lithium and decrease potassium). On occasion, some patients complain of decreased cognitive functions. After a few weeks of treatment, some of the adverse effects (especially thirst and tremor) tend to disappear. If the patient has complaints about gastro-intestinal problems, switching to another lithium preparation may be called for. Another possible malady is hypothyroidism and is usually treated with supplemental thyroid hormone (Ambrosiani et al., 2018). Hyperparathyroidism (parathyroid glands create high amounts of parathyroid hormone in the bloodstream) and consequent hypercalcemia (calcium level in blood is above normal) can also arise during long-term treatment with lithium.

Patients who have or have had acute myocardial infarction, acute kidney failure, or certain rare disorders of heart rhythm (notably, Brugada syndrome of ventricular arrhythmia—electrical activity of heart is abnormal). With close cautious medical monitoring lithium can be used in the presence of cardiac arrhythmia, reduced kidney function, psoriasis, myeloid leukemia, Addison's disease, hypothyroidism, and certain neurological disorders, including abnormalities of posture and movement, myasthenia gravis—weak, fatigued muscles under voluntary control, and epilepsy. Syndrome of Irreversible Lithium-Effectuated Neurotoxicity or the acronym SILENT is a rare, probably irreversible is persistent cerebellar dysfunction with ataxia or unsteady gait that has been reported in a few instances (Aditanjee et al., 2005).

Lithium dosage for older patients is to be determined by the attending clinician; typically the lithium dose for older patients should be about 20% lower those for younger patients (Shulman et al., 2019). In addition to the adverse effects in the elderly, already described common adverse effects can include confusion or worsening of cognitive functions, unsteady balance and gait (ataxia), restless movements (akathisia), declining kidney function, hypothyroidism, possible worsening of diabetes mellitus, and leg swelling (peripheral edema).

Increased risk occurs whenever abnormally high blood levels of lithium (above 1.5 mEq/L) are present in a patient. This situation can increase the risk of intoxication, signs of which are severe tremor, confusion, vomiting, abdominal pain, diarrhea, abnormally increased reflexes, speech difficulties, abnormal heart rhythm, hypotension, and convulsions. If such symptoms occur, the prescribing clinician or an emergency service should be consulted immediately. To determine the level of lithium in the blood, it is required to test the blood; if it is very high (above 2 mEq/L), blood dialysis may be needed to remove lithium—this is a decision that should be made by a kidney specialist. Note that even if measured blood levels are not greatly elevated, intoxication may be present due to circulating levels in blood not being in equilibrium with those in the CNS, and lithium should be stopped as clinical evaluation is pursued (Tondo et al., 2019).

A MATTER OF LONG-TERM DISCUSSION

One facet in the use of lithium that has undergone long-term and on-going discussion is its effect in pregnancy and post-partum. In the past, the mainstream opinion was that lithium should be discontinued immediately when pregnancy was ascertained or planned. The major concern was risk of a rare congenital cardiac abnormality (Ebstein's anomaly, arising in approximately 1 in 20,000 live births without lithium exposure—the anomaly is a complex congenital heart defect characterized by a malformation of the tricuspid valve and right side of the heart). Note that recent studies have confirmed the association between maternal exposure to lithium in the first 3 months of pregnancy and increased risk of this and other fetal malformations but their risk-estimates are lower than previously reported (Patorno et al., 2017; Munk-Olsen et al., 2018; Poels et al., 2018). During the first trimester, reduction of lithium doses may be considered but weighed against especially the high threats of relapse (especially of depressive or mixed states of bipolar disorder) early in pregnancy connected with treatment discontinuation (Viguera et al., 2007). Whenever lithium is prescribed during pregnancy, serum concentrations should be monitored frequently (at least monthly, and preferably weekly in the third trimester) (Wesseloo et al., 2017). There is no need to stop lithium before delivery, but blood levels should be measured twice weekly for the first 2 weeks after delivery (Wesseloo et al., 2017). When delivery is prolonged, it is vital to ensure a good fluid intake during treatment with lithium. It is recommended that clinicians and patients should discuss risks and benefits of continuing, lowering, or interrupting lithium treatment in anticipation of pregnancy and after childbirth—this includes during breast-feeding (Poels et al., 2018).

DID YOU KNOW?

Note that the mechanism of action of lithium is unknown. It is rapidly absorbed, has a relatively small volume of distribution, and leaves the body in urine which is unchanged because the human body does not metabolized lithium.

What about post-partum treatment?

In a pregnant woman with bipolar disorder, when and if lithium is discontinued, it should be restarted immediately after delivery. This is the case because the early post-partum period carries a high risk for recurrences of bipolar disorder and depression (Viguera et al., 2007; Wesseloo et al., 2016). After childbirth and during the first month after childbirth, serum lithium levels are often increased to 0.8–1.0 mEq/L to minimize risk of illness; they also checked at least weekly for a month. In a case where the blood level is stable, it can be checked monthly for up to 6 months. During lithium treatment, breast-feeding is not recommended (because concentrations in breast milk are about half of the concentrations in maternal blood); moreover, infant feeding formula should be used (Galbally et al., 2018; Poels et al., 2018).

The bottom line: when the time comes to limit intake of lithium, this should be done gradually, and if the treatment needs to be rapidly discontinued, this should only be carried out under close medical supervision (Tondo et al., 2019).

DID YOU KNOW?

It is no secret that cocaine was an ingredient of the soft drink Coca Cola in the late-19th and early-20th centuries, what is not as well known is that lithium citrate was an ingredient in what is now known as 7 Up. However, in 1948, all beverage makers were forced to remove lithium.

THE 411 ON META-ANALYSIS

Note that a large amount of research conducted on lithium and its uses in clinical treatment, its efficacy in treatment, and its pros and cons and other important and findings via meta-analysis. For those not familiar with meta-analysis what it boils down to is a method for statistical analysis data which are made available to clinicians and researchers. A meta-analysis is usually preceded by a systematic review, as this allows identification and critical appraisal of all the relevant evidence which works to limit the risk of bias in summary estimates. Riley et al. (2010) point out that these meta-analysis have become increasingly popular. The general steps are then as follows: (Herrera Ortiz et al., 2022).

1. Formulation of the research question, e.g., using the PICO(T) model (Population, Intervention, Comparison, Outcome, Time Period)
2. Search of literature
3. Incorporation criteria—selection of studies
    1. Based on quality, e.g., the requirement of randomization and blind in a clinical trial
    2. Selection of specific studies on a well-specified subject, e.g., the treatment of prostate cancer,
    3. Determine whether unpublished studies are included to avoid *publication bias* (aka *file drawer problem*—this occurs when the outcome of an experiment or research study biases the decision to publish or otherwise distribute it. Publishing only results that show a significant finding disturbs the balance of finding and disturbs the balance of findings in favor of positive results (Song et al., 2010).
    4. Determine which dependent variables or summary measures are allowed. For example, when thinking about a meta-analysis of published (cumulative) data:
        – Differences (discrete data)
        – Means (continuous data)
        – Hedges' g is a popular summary measure for continuous data that is standardized in order to eliminate scale differences—basically, measures effect size and helps determine how much two groups differ:

$$\Delta = \frac{\mu_t - \mu_c}{\sigma}$$

    where
    μ_t is the treatment mean.
    μ_c is the control mean.
    σ^2 is the pooled variance.
5. Choice of a meta-analysis model
6. Scrutinize sources of between-study heterogeneity

When performing a meta-analysis, two types of evidence can be singled out: aggregate data (AD) (either direct or indirect) and individual participant data (IPD).

IPD is raw data from individual participants and is often used in meta-analysis; it is considered the gold standard of evidence synthesis (Thomas et al., 2014). This superior ranking is attributed to the IPD's high level of precision and consistency; this approach makes it easier for researchers to minimize heterogeneity.

Simply stated, AD is a combination of individual data (see Figure 11.1). The big difference or distinction between AD and IPD. As stated earlier, individual data are disaggregated individual results and are used to conduct analyses for estimation of subgroup difference. On the other hand, IPD disaggregated individual results and are used to conduct analyses for estimation subgroup differences (Jacob, 2016). Note that AD data are primarily used by researchers and analysts, banks, administrators, and policymakers for varied reasons. Also note that AD is widely used, but it also has limitations, including drawing inaccurate inferences and false conclusions which are also termed "ecological fallacy" (Starrin et al., 1993); this term was coined by Selvin (1958). An ecological fallacy (also population fallacy or ecological inference fallacy (Ess and Sudweeks, 2001) is a formal fallacy in the interpretation of statistical data that occurs when inferences about the nature of individuals are deduced form inference about the group to which those individuals belong.

It is important to point out that all legitimate medical analysis is subject to review—peer review. These analyses receive systematic reviews that were and are developed out of a need to ensure that decisions affecting people's lives can be informed by an up-to-date and complete understanding of the relevant research evidence (Cochran Handbook, 2019).

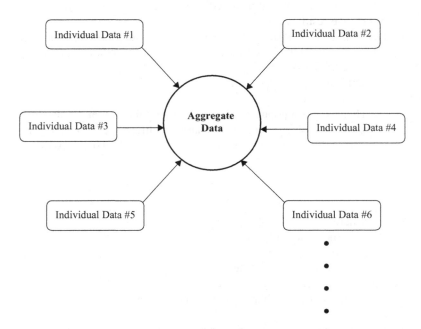

FIGURE 11.1 The basic meaning of aggregate data is a combination of individual data.

The bottom line on meta-analysis (or any other type of medical research and analysis) is to keep in mind that most medical research and analysis provides needed information and new findings that are sometimes revolutionary in content and certainly needed in the treatment of patients suffering from mania or bipolar disorders and those additional disorders that are listed and discussed in the following chapter. Not all the "news" is good news, however. There are researchers who make breakthrough findings that are important but not necessarily good news (e.g., dangerous or alarming side effects). And it is important to consider bias and researchers' bias. Consider, for example, the researcher who discovers something new and important in medical treatment of patients with various disorders. In the researcher's research, he or she finds some good and something not so good (side effects, or worse). This researcher wants his or her findings published in reputable medical journals and maybe even television appearances and selected as a keynote speaker at highly reputable medical journal conferences and others. Well herein lies the dilemma; that is, the researcher might be tempted to only discuss the positive (good news) findings because he or she has an ego that desires appreciation. In the publishing world, it is so much easier to be recognized and published if your message is positive, highlighting the so-called good news and ignoring the limitations or worse. When publishing or other disclosure only points out results that show a significant finding is real and it disturbs the balance of findings—this is known as publication bias. And it exists; it is real.

NOTE

1 Much of the information that follows is based on and adapted from Tondo et al. (2019).

REFERENCES

AGAH (2004). Pharmacokinetics. Accessed 11/22/22 @ https://www.agah.eu/.

Abou-Saleh, M.I., Muller-Oerlinghausen, B., and Coopen, A.J. (2017). Lithium in the episode and suicide prophylaxis and in augmenting strategies in patients with unipolar depression. *International Journal of Bipolar Disorders*, v. 5, p. 11.

Aditanjee, D., Munsh, K.R., and Thampy, A. (2005). The syndrome of irreversible lithium-effectuated neurotoxicity. *Clinical Neuropharmacology*, v. 28, pp. 38–49.

Aherns, B., and Muller-Oerlinghausen, B. (2001). Does lithium exert an independent antisuicidal effect? *Pharmacopsychiatry*, v. 34, pp. 132–136.

Alevizos, B., Alevizos, E., Leonardou, A., and Zervas, I. (2012). Low dosage lithium augmentation in venlafaxine resistant depression: open-label study. *Psychiatrike*, v. 23, pp. 143–148.

Ambrosiani, L., Pisanu C., Deidda, A., Chilloti, C., Stochino, M.E., and Rocchetta, A. (2018). Thyroid and renal tumors in patients treated with long-term lithium. *International Journal of Bipolar Disorders*, v. 6, pp. 17–28.

Amdisen, A. (1967). Serum lithium determinations for clinical use. *Scandinavian Journal of Clinical and Laboratory Investigation*, v. 20, pp. 104–108.

American Psychiatric Association (2022). *Diagnostic and statistical manual of mental disorders* (5th ed.). New York: APA.

Angst, J., Weis, R. Grof, P., Baastrup, P.C., and Schou, M. (1970). Lithium prophylaxis in recurrent affective disorders. *British Journal of Psychiatry*, v. 116, pp. 604–614.

Angst, J., Angst, F., Gerber-Werder, R., Gamma, A. (2005). Suicide in 406 mood-disorder patients with and without long-term medication: A 40–44-year follow-up. *Archives of Suicide Research*, v. 9, pp. 279–300.

Austin, M.P., Souza, F.G., and Goodwin, G.M. (1991). Lithium augmentation in antidepressant-resistant patients: Quantitative analysis. *British Journal of Psychiatry*, v. 139, pp. 510–514.

Baastrup, P.C. (1964). The use of lithium in manic-depressive psychosis. *Comprehensive Psychiatry*, v. 5, pp. 394–408.

Baastrup, P.C., and Schou, M. (1968). Prophylactic lithium. *Lancet*, v. 1, pp. 419–422.

Baastrup, P.C., Paulsen, I.C., Schou, M., Thomsen, K., and Amdisen, A. (1970). Prophylactic lithium: Double blind discontinuation in manic-depressive and recurrent-depressive disorders. *Lancet*, v. 2, no. 7668, pp. 326–330.

Baethge, C., Baldessarini, R.J., Bratti, I.M., and Tondo, L. (2003). Prophylaxis-latency and outcome in bipolar disorders. *The Canadian Journal of Psychiatry*, v. 48, pp. 449–457.

Baldessarini, R.J., Tondo, L., Hennen, J., and Viguera, A.C. (2002). Is lithium still worth using? An update of selected recent research. *Harvard Review of Psychiatry*, v. 10, pp. 59–78.

Baldessarini, R.J., Leahy, L.F., Arcona, S. Gause, D. Zhang, W., and Hennen, J. (2007). Prescribing patterns of psychotropic medicines in the United States for patients diagnosed with bipolar disorders. *Psychiatric Services*, v. 58, pp. 85–91.

Baldessarini, R.J., Henk, H.J., Sklar, L.R., Chang, J., and Leahy, L.F. (2008). Psychotropic medications for bipolar disorder in the United States polytherapy and adherence. *Psychiatric Services*, v. 59, pp. 1175–1183.

Baldessarini, R.J., and Tondo, L. (2009). Suicidal risks during treatment of bipolar disorder patients with lithium vs. anticonvulsants. *Pharmacopsychiatry*, v. 4, pp. 72–75.

Baldessarini, R.J., Undurraoa, I., Vasquez, G., Tondo, L, Salvatore, P., Ha, K. et al. (2012). Predominant recurrence polarity among 929 adult international bipolar disorder patients. *Acta Psychiatrica Scandinavica*, v. 125, pp. 292–302.

Baldessarini, R.J. (2013). *Chemotherapy in psychiatry* (3rd ed.). New York: Springer Press.

Baldessarini, R.J., and Tondo, L. (2018). Effects of treatment discontinuation in clinical psychopharmacology. *Psychotherapy and Psychosomatics*, v. 88, pp. 65–70.

Baldessarini, R.J., Tondo, L., and Vasquez, G. (2019). Unmet needs in psychiatry: Bipolar depression, in Pompili, M., McIntyre, R.S., Fiorillo, A., and Sartorius, N., eds. *New directions in psychiatry*. New York: Springer Press.

Bauer, M., and Dopfmer, S. (1999). Lithium augmentation in treatment-resistant depression meta-analysis of placebo-controlled studies. *Journal of Clinical Psychopharmacology*, v. 9, pp. 427–434.

Bauer, M., Forsthoff, A., Baethge, C., Adli, M., Berghofer, A., Dopfmer, S, et al. (2003). Lithium augmentation therapy in refractory depression-update. *European Archives of Psychiatry and Clinical Neuroscience*, v. 253, pp. 132–139.

Bauer, M., and Gitlin, M. (2016). *The essential guide to lithium treatment*. Basel: Springer International Press.Berghofer, A., Adli, M., Baethge, C. Bauer, M., Bschor, T., et al. (2008). Long-term effectiveness of lithium in bipolar disorder: Multicenter investigation of patients with typical and atypical features. *The Journal of Clinical Psychiatry*, v. 69, pp. 1860–1868.

Blackwell, B., and Shepherd, M. (1968). Prophylactic lithium: Another therapeutic myth? An examination of the evidence to date. *Lancet*, v. 291, no. 7549, pp. 968–971.

Bratti, L.M., Baldessarini, R.I., Baethge, C., and Tondo, L. (2003). Pretreatment episode count and response to lithium treatment in manic-depressive illness. *Harvard Review of Psychiatry*, v. 11, pp. 245–256.

Cade, J.F.J. (1949). Lithium salts in the treatment of psychotic excitement. *Medical Journal of Australia*, v. 2, no. 36, pp. 349–352.

Cade, J.F.J. (1999). John Fredrick Joseph Cade: Family memories on the occasion of the 50th anniversary of the discovery of the use of lithium in mania. *The Australian and New Zealand Journal of Psychiatry*, v. 33, no. 5, pp. 615–618.

Cipriani, A., Pretty, H., Hawton, K., and Geddes, J.R. (2005). Lithium in the prevention of suicidal behavior and all-cause mortality in patients with mood disorders: Systematic review of randomized trial. *American Journal of Psychiatry*, v. 162, pp. 1805–1819.

Cipriani, A., Hawton, K., Stockton, S., and Geddes, J.F. (2013). Lithium in the prevention of suicide in mood disorders: Updated systematic review and meta-analysis. *BMJ*, v. 246, pp. 3546–3558.

Cochran Handbook (2019). *Cochran Handbook for systematic review of intervention* (2nd ed.). New York: Wiley-Blackwell.

Coppen, A., Abou-Saleh, P., Milln, P., Baily, J., and Wood, K. (1983). Decreasing lithium dosage reduces morbidity and side-effects during prophylaxis. *Journal of Affection Disorders*, v. 5, pp. 353–362.

Dusetzina, S.B., et al. (2012). Treatment use and costs among privately insured youths with diagnosis with bipolar disorder. *Psychiatric Services*, v. 63, pp. 1019–1025.

el-Mallakn, R.S. (1990). Lithium. *Connecticut Medicine*, v. 54, no. 3, pp. 115–126.

Ess, C., and Sudweeks, F. (2001). *Culture, technology, communication: Towards an intercultural global village.* New York: SUNY Press.

Faedda, G.L., Tondo, L., Baldessarini, R. Suppes, T., and Tohen, M. (1993). Outcome after rapid vs. gradual discontinuation of lithium treatment in bipolar mood disorders. *Archives of General Psychiatry*, v. 50, pp. 448–455.

Forte, A., Baldessarini, R.J., Tondo, L., Vasquez, G. Pompili, M., and Giradi, P. (2015). Long-term morbidity in bipolar I and bipolar II, and major depressive disorders. *Journal of Affective Disorders*, v. 178, pp. 71–78.

Galbally, M., Bergink, V., Vigod, S., Buist, A. Boyce, R., et al. (2018). Is breast always best? Breasting feeding and lithium. *Lancet Psychiatry*, v. 5, pp. 534–536.

Geddes, J.R., Burgess, S., Hawton, K. Jamison, K., and Goodwin, G.M. (2004). Long-term lithium therapy for bipolar disorder systematic review and meta-analysis of randomized controlled trials. *American Journal of Psychiatry*, v. 161, pp. 217–222.

Geddes, J.R., Goodwin, G.M. Rendell, J., Azorin, J.M., Cipriani, A., and Ostacher, M.J. (2010). Lithium plus valproate combination therapy vs. monotherapy for relapse prevention in bipolar I disorder: Randomized, open-label trial. *Lancet*, v. 375, p. 395.

Goodwin, F., Fireman B., Simon, G.E., Hunkeler, E.M., Lee, J., and Revick, D. (2003). Suicide risk in bipolar disorder during treatment with lithium and divalproex. *JAMA*, v. 290, pp. 1467–1473.

Grof, P. (2006). *Lithium in neuropsychiatry: The comprehensive guide.* London: Informa, pp. 157–178.

Grof, P., Schou, M., Angst, J., Baastrup, P.C., and Weis, P. (1970). Methodological problems of prophylactic trial in recurrent affective disorders. *British Journal of Psychiatry*, v. 116, pp. 599–603.

Guzzetta, F., Tondo, L. Centorrino, F., and Baldessarini, R.J. (2007). Lithium treatment reduces suicide risk in recurrent major depressive disorder. *Journal of Clinical Psychiatry*, v. 68, pp. 380–383.

Harris, E.C., and Barraclough, B. (1997). Suicide as an outcome for mental disorders: Meta-analysis. *British Journal of Psychiatry*, v. 170, pp. 205–228.

Haussmann, R., Lewitzka U., Severus, E., and Bauer, M. (2017). Correct treatment of mood disorders with lithium (German). *Der Nervenarzt*, v. 88, pp. 1323–1334.

Herrera Ortiz, A.F., Cadavid Camacho, E., Cubillos, Rojas, J., Cavavid Camacho, T., Zoe Guevara, S., Tatiana Rincon Cuenca, N., Vasquez Perdomo A., Del Castillo Herazo, V., and Giraldo Male, R., (2022). A Practical guide to perform a systematic literature review and meta-analysis. *Principles and Practice of Clinical Research*, v. 7, no. 4, pp. 47–57.

Huyse, F.J., Touw D.J., van Schijndel, R.S., de Lange, J.J. Slaets, J.P. (2006). Psychotropic drugs and the perioperative proposal for a guideline in elective surgery. *Psychosomatics*, v. 47, pp. 8–72.

Jacob, R. (2016). Using aggregate administrative data in social policy research. Accessed 12/12/22 @ https://www.act.hhs.gov/epre/resource/using-aggregate-administrative-dat-in-soicla-policy-research.

Johnson, G., and Gershon, S. (1999). Early North American research on lithium. *Australian and New Zealand Journal of Psychiatry*, v. 33, pp. s48–s53.

Kazdin, A.E. (ed,). (2000). *Encyclopedia of psychology*. Washington, DC: APA.

Kessing, L.V., Sondergard, L., Kvist, K., and Andersen, P.K. (2005). Suicide risk in patients treated with lithium. *Archives of General Psychiatry*, v. 62, pp. 860–866.

Kessing, L.V., Gerds, T.A., Knudsen, N.N., Jorgensen, L.F., Kristiansen, S.M., and Voutchkova, D., et al. (2017). Association of lithium in drinking water with the incidence of dementia. *JAMA Psychiatry*, v. 74, pp. 1005–1010.

Kleindienst, N., and Greil, W. (2000). Differential efficacy of lithium and carbamazepine in the prophylaxis of bipolar disorder: Results of the MAP study. *Neuropsychobiology*, v. 42, no. Suppl 1, pp. 2–10.

Kukopulos, A., Reginaldi, D., Laddomada, P., Floris, G., Serra, G., and Tondo, L. (1980). Course of the manic-depressive cycle and changes caused by treatment. *Pharmacopsychiatry Neuropsychopharmacology*, v. 13, pp. 156–167.

Kukopulos, A., Reginaldi, D, Tondo, L., Visioli, C., and Baldessarini, R.J. (2013). Course sequences in bipolar disorder: Depressions preceding or following manias or hypomanias. *Journal of Affective Disorders*, v. 151, pp. 105–110.

Lauterbach, F., Felber, W., Muller-Oerlinghausen, B., Ahrens, B., Bronisch, T., Meyer, T. et al., 2008. Adjunctive lithium treatment in the prevention of suicidal behavior in depressive disorders: randomized, placebo-controlled, 1-year trial. *Acta Psychiatrica Scandinavica*, v. 118, pp. 469–479.

Malhi, G.S., Gessler, D., and Outhred, T. (2016). Use of lithium for the treatment of bipolar disorder: Recommendations from clinical practice guidelines. *Journal of Affective Disorders*, v. 217, pp. 266–280.

Manchia, M., Hajeck, T., O'Donovan, C., Deiana, V., Chillotti, C., Ruzickova, M., et al. (2013). Genetic risk of suicidal behavior in bipolar spectrum disorder: Analysis of 737 pedigrees. *Bipolar Disorder*, v. 15, pp. 496–506.

McKnight, R.F., Adida, M., Budge, K., Stockton, S., Goodwin, G.M., and Geddes, J.R. (2012). Lithium toxicity profile: Systematic review and meta-analysis. *Lancet*, v. 379, no. 9817, pp. 721–728.

Mentalhelp (2022). Understanding mood episodes in depression. Accessed 11/18/22 @ https://www.mentalhelp.net/depression/.

Miura, T., Noma, H., Furukawa, T.A., Mitsuyasu, H., Tanaka, S., Stockton, S., et al. (2014). Comparative efficacy and tolerability of pharmacological treatments in the maintenance treatment of bipolar disorder a systematic review and network meta-analysis. *Lancet Psychiatry*, v. 1, pp. 351–359.

Muller-Oerlinghausen, B., Muser Causemann, B., and Volk, J. (1992). Suicides and parasuicides in a high-risk patient group on and off lithium long-term medication. *J Affect Disord*. 25:261–70.

Muller-Oerlinghausen, B., Ahrens, B., and Felber, W. (2006). Suicide-preventive and mortality-reducing effect of lithium, in *Lithium in neuropsychiatry*. London: Informa Healthcare, pp. 179–192.

Munk-Olsen, T., Liu, X., Viktorin, A. Brown, H.K., Di Florio, A., D'Onofrio, B.M., et al. (2018). Maternal and infant outcomes associated with lithium use in pregnancy. An international collaboration combining data from 6 cohort studies using meta-analysis covering 727 lithium exposed pregnancies and 21,397 bipolar or major depressive disorder reference pregnancies. *Lancet Psychiatry*, 5L644–53.

Neilson, R., Kessing, L., Nolen, J., and Licht, R.W. (2018). Lithium and renal impairment: A review on a still hot topic. *Pharmacopsychiatry*, v. 51, no. 15, pp. 200–205.

Osby, U., Brandt, L., Correia, N., Ekbom, A., and Sparen, P. (2001). Excess mortality in bipolar and unipolar disorder in Sweden. *Archives of General Psychiatry*, v. 58, pp. 844–850.

Pacchiarotti, I., Bond, D.J., Baldessarini, R.I., Nolen, W.A., Grunze, H., Licht, R.W., et al. (2013). International Society for Bipolar Disorders (ISBD) task-force report on antidepressant use in bipolar disorders. *American Journal of Psychiatry*, v. 170, pp. 1249–1262.

Patel, N., Viguera, A.C., and Baldessarini, R.J. (2018). Mood stabilizing anticonvulsants, spina bifida, and folate supplementation. *Journal of Clinical Psychopharmacology*, v. 39, pp. 7–10.

Patorno, F., Huybrechts, K.F., Bateman, B.T., Cohen, J.M. Desai, R. J., Mogun H., et al. (2017). Pregnancy and the risk of cardiac malformations. *New England Journal of Medicine*, v. 376, pp. 2245–2254.

Perugi, G., Sani, G., and Tondo, L. (2019). *Practical guide to the use of lithium in the treatment of bipolar disorder patients* (Italian). Rome: About Books.

Poels, E.M.P., Bijma, H.H., Galbally, M., and Bergink, V. (2018). Lithium during pregnancy and after delivery: Review. *International Journal of Bipolar Disorders*, v. 6, pp. 26–37.

Poon, S.H., Sim, K., Sum, M.Y., Kuswanto, C.N., and Baldessarini, R.J. (2021). Evidence-based options for treatment-resistant adult bipolar disorder patients. *Bipolar Disorders*, v. 14, pp. 573–584.

Popovic, D., Reinares, M., Goikolea, J.M., Bonnin, C.M., Gonzalex-Pinto, A., and Vieta, E. (2012). Polarity index of bipolar disorder. *European Neuropsychopharmacology*, v. 22, pp. 339–346.

Post, R., M., Leverich, G.S., Kupka, R.W., Keck, P.E. Jr., McElroy, S.L., Altshuler, L.L., et al. (2010). Early-onset bipolar disorder and treatment delay are risk factors for outcome in adulthood. *Journal of Clinical Psychiatry*, v. 71, pp. 864–872.

Riley, R.D., Lambert, P.C., and Abo-Zaid, G. (2010). Meta-analysis of individual participant data: Rational, conduct, and reporting. *BMJ*, v. 340, p. c221.

Selvin, H.C. (1958). Durkheim's suicide and problems of empirical research. *American Journal of Sociology*, v. 63, no. 6, pp. 607–619.

Sharma, R., and Markar, H.R. (1994). Mortality in affective disorder. *Journal of Affective Disorders*, v. 31, pp. 91–96.

Shulman, K.L., Almeida, O.P., Herrmann, N., Schaffer, A., Strejilevich, S.A., Paternoster, C. et al. (2019). Delphi survey of maintenance lithium treatment in older adults with bipolar disorder: An ISBD task force report. *Bipolar Disorders*, v. 21, pp. 117–123.

Smith, E.G., Sondergard, L., Lopez, A.G., Andersen, P.K., and Kessing, L.V. (2009). Association between consistent purchase of anticonvulsants or lithium and suicide risk: Longitudinal cohort study from Denmark. *Journal of Affective Disorders*, v. 117, pp. 162–167.

Snow, F., Parekh, S., Hooper, L., Loke, Y.K., Ryder, J, Sutton, A. J., Hing, C., Kwok, C.S., Pang, C., and Harvey, I. (2020). Dissemination and publication of research findings: An updated review of related biases. *Health Technology Assessment*, v. 14, no. 8, pp. 1–220.

Song, Y., et al. (2010). Neural and synaptic defects in Slytherin, a Zebroids model for human congenital disorders of glycosylation. *PLoS One*, v. 5, no. 10, p. e13743.

Starrin, B., Hagquist, C., Larsson, G., and Svensson, P.-G. (1993). Community types, socioeconomic structure and IHD mortality—A contextual analysis based on Swedish aggregate data. *Social Science & Medicine*, v. 36, no. 12, pp. 1569–1578.

Staudt-Hansen, P., Frahm Laursen, M., Grontved, S., Puggard Vogt Staszek, S., Licht, R.W., and Nielsen, R.N. (2019). Increasing mortality gap for patients diagnosed with bipolar disorder—A nationwide study with 20 years of follow-up. *Bipolar Disorders*, v. 31, pp. 270–275.

Thies-Flechtner, K., Muller-Oerlinghausen, B., Seibert, W., Walther, A., and Greil, W. (1996). Effect of prophylactic treatment on suicide risk in patients with major affective disorders: Data from a randomized prospective trial. *Pharmacopsychiatry*, v. 29, pp. 103–107.

Thomas, D., Radji, S., and Benedetti, A. (2014). Systemic review of methods for individual patient data meta-analysis with binary outcomes. *BMC Medical Research Methodology*, v. 14, no. 1, p. 79.

Tiihonen, J., Tanskanen, A. Hoti, F. Vattulainen, P., Taipale, H., Mehtala, J., et al. (2017). Pharmacological treatments and risk of readmission to hospital for unipolar depression in Finland: A nationwide cohort study. *Lancet Psychiatry*, v. 4, pp. 547–558.

Tondo, L., Baldessarini, R.J., Hennen, J., Floris, G., Silvetti, F., and Tohen, M. (1998). Lithium treatment and risk of suicidal behavior in bipolar disorder patients. *Journal of Clinical Psychiatry*, v. 59, pp. 405–414.

Tondo, L., Hennen, J., and Baldessarini, R.J. (2001). Reduced suicide risk with long-term lithium treatment in major affective illness: Meta-analysis. *Acta Psychiatrica Scandinavica*, v. 104, pp. 163–172.

Tondo, L., Hennen, J., and Baldessarini, R.J. (2003). Rapid-cycling bipolar disorder: Effects of long-term treatments. *Acta Psychiatrica Scandinavica*, v. 108, pp. 4–14.

Tondo, L., and Baldessarini, R.J. (2016). Suicidal behavior in mood disorders: Response to pharmacological treatment. *Current Psychiatry reports*, v. 10, pp. 88–98.

Tondo, L. Abramowicz, M., Alda, M. Bauer, M., Bocchetta, A. Bolzani, L., et al. (2017). Long-term lithium treatment in bipolar disorder: Effects on glomerular filtration rate and other metabolic parameters. *Journal of Bipolar Disorders*, v. 5, p. 27.

Tondo, L. et al. (2019). Clinical use of lithium slats: Guide for users and prescribers. *International Journal of Bipolar Disorders*, v. 7, p. 16. Accessed 11/17/22 @ https://doi.org/1086/s40345-019-0151-2.

Tsujii, T., Uchida, T., Suzuki, T., Mimura, m., Hirano, J., and Uchida, H. (2019). Factors associated with delirium following electroconvulsive therapy. A systematic review, *Journal of ECT*, v. 35, no. 4, pp. 279–287.

Undurraga, J., Tondo, L., Gorodischer, A., Azua, F., Tay, K.H., et al. (2019). Lithium treatment for unipolar major depressive disorder: Systematic review. *Journal of Psychopharmacology*, v. 16, pp. 167–176.

Vasquez, G.H., Tondo, L., Undurraga, J., and Baldessarini, R.J. (2013). Overview of antidepressant treatment in bipolar depression: Critical commentary. *International Journal of Neuropsychopharmacology*, v. 16, pp. 1673–1685.

Vieta, E., Berk, M., Wang, W., Colom, F., Tohen, M., and Baldessarini, B.J. (2009). Predominant previous polarity as an outcome predictor in a controlled treatment trial for depression in bipolar I disorder patients. *Journal of Affective Disorders*, v. 19, pp. 22–27.

Vieta, E., Gunther, O., Locklear, J., Ekman, M., Miltenburger, C., Chatterton, C., et al. (2011). Effectiveness of psychotropic medications in the maintenance phase of bipolar disorder: A meta-analysis of randomized controlled trials. *International Journal of Neuropsychopharmacology*, v. 14, pp. 1029–1049.

Viguera, A.C., Whitfield, T., Baldessarini, R.J., Newport, D., Stowe, Z., and Cohen, L.S., (2007). Recurrences of bipolar disorder in pregnancy: prospective study of mood-stabilizer discontinuation. *American Journal of Psychiatry*, v. 164, pp. 1817–1824.

Wesseloo, R., Kamperman, A., Munk-Olsen, T., Pop, V., Kushner, S.A., and Bergink, V. (2016). Postpartum episodes in women at high risk, system review and meta-analysis. *American Journal of Psychiatry*, v. 173, pp. 119–127.

Wesseloo, R., Wierdsma, A., Hoogendijk, W.J., Munk-Olsen, T., Kushner, S.A., and Bergink, V. (2017). Lithium dosing during pregnancy. *British Journal of Psychiatry*, v. 211, pp. 31–36.

Yatham, L. N., Kennedy, S.H., Parikn, S.V., Schaffer, A., Bond, D.J., Frey R.N., et al. (2018). Canadian Network for Mood and Anxiety Treatments (CANMAT) and international Society for Bipolar Disorders (ISBD) 2018 Guidelines for management of patients with bipolar disorder. *Bipolar Disorders*, v. 20, pp. 97–170.

12 Medical Lithium
Adverse Effects

INTRODUCTION

The preceding chapter presented information about lithium's use in treating mania, bipolar, and depression disorders. Lithium is also used in treating cluster headaches, migraine, and hypnic headache (i.e., primary headaches that affect the elderly). Along with the good results of lithium treatment, there are or can be side effects that are more adverse than good.

ADVERSE EFFECTS OF LITHIUM TREATMENT

Common adverse effects of lithium include [sources: DuraPoint® System (2013); Australian Medicines Handbook (2013); Joint Formulary Committee (2013); Lithium (Rx) (2013); National Library of Medicine (2013); Aiff, et al., (2015); TGA eBusiness Services (2017)]:

- Acne
- Confusion
- Constipation (usually transient, but can be persistent)
- Memory loss and decreased memory
- Diarrhea (usually transient, but can persist in some)
- Dry mouth
- EKG changes—usually benign changes in T waves
- Euthyroid goitre or goiter—i.e., the formation of a goiter despite normal thyroid functioning.
- Extrapyramidal side effects—movement-related problems such as muscle rigidity, parkinsonism, and dystonia (i.e., abnormal postures as a result of sustained muscle contractions)
- Hand tremor—(usually transient but can persist in some). If severe, psychiatrist may lower lithium dosage, change lithium salt type or modify lithium preparation from long to short acting (despite lacking evidence for these procedures) or use pharmacological help (Baek et al., 2013).
- Hair loss/hair thinning
- Hyperreflexia—reflexes that are over responsive
- Hypoglycemia (Malhi et al., 2013)
- Hypothyroidism—thyroid hormone deficiency
- Leukocytosis—white blood cell count is elevated
- Muscle weakness (usually transient but can persist in some)
- Myoclonus—muscle twitching

- Nausea—(Usually transient; Gitlin, 2021)
- Polydipsia—increased thirst (can also be caused by other medications such as anticholinergics, demeclocycline, diuretics, phenothiazines)
- Polyuria—increased urination
- Renal (kidney) toxicity which could lead to chronic kidney failure
- Vomiting (usually transient but can persist in some)
- Vertigo
- Weight gain—lithium carbonate can induce 1–2 kg of weight gain (Malhi et al., 2013). Weight gain may be a source of low self-esteem for the clinically depressed (Sperner-Unterweger et al., 2001).

DID YOU KNOW?

The majority of side effects of lithium are dose-dependent. The lowest effective dose is used to control the risk of side effects.

HYPOTHYROIDISM

Hypothyroidism is a condition marked by having an underactive thyroid. It is a glandular disorder where the body does not produce enough thyroid hormones to meet the body's needs. Note that there is general agreement that hypothyroidism is a possible consequence of lithium treatment but there is little agreement as to what the incidence of lithium-related hypothyroidism is, or what clinical and laboratory tests could best be used to evaluate thyroid function (Piziak et al., 1978). It is important to point out that lithium can cause hyperthyroidism as well as hypothyroidism. Moreover, symptoms of hyperthyroidism can mimic those of mania (Fairbrother et al., 2019).

SEROTONIN SYNDROME

Serotonin syndrome is found among people who are treated with lithium carbonate; it manifests itself especially in people who are female, 60+, have been taking the drug for <1 month. People receiving lithium treatment and are suffering with any of the following symptoms should seek medical help immediately.
   Symptoms:

- tachycardia of rapid heartbeat
- hallucinations
- anxiety
- delirium
- excessive sweating
- muscle rigidity
- dilated pupils

So, what exactly is serotonin syndrome?

Well, first of all, be aware that serotonin syndrome is a potentially life-threatening condition precipitated by the use of serotonergic drugs. These drugs affect the neurotransmitter serotonin. Serotonin syndrome may be a result of therapeutic medication use, interactions between medications or recreational drugs, or intentional overdose. In addition to the symptoms listed above, other symptoms can range from mild to fatal and classically include altered mental status, autonomic dysfunction, and neuromuscular excitation (Simon and Keenaghan, 2022). There are several criteria (aka decision rules) for making this clinical diagnosis, but the Hunter criteria are generally accepted as the most accurate. Note that Hunter criteria requires spontaneous clonus, inducible clonus, ocular clonus, agitation, diaphoresis, tremor and hyperreflexia for accurately predicting serotonin toxicity. Clonus may be conceptualized as a type of profound hyperreflexia, in which each muscle contraction triggers another reflexive contraction. In other words, clonus is a set of involuntary and rhythmic muscular and contractions and relaxations (Hilder and Rymer, 1999). Management consists of immediate discontinuation of serotonergic agents, hydration, and supportive care to manage blood pressure, hyperpyrexia, and respiratory and cardiac complications (Francescangeli et al., 2019; Duma and Fung, 2019; Srivastava et al., 2019).

RENAL SIDE EFFECTS

While it is true that lithium has been an essential treatment for bipolar disorder for almost 60 years, it is also true that clinical use of lithium has been problematic due to its narrow therapeutic index (TI). As pointed out earlier the TI (aka *therapeutic ratio*) is a ratio that compares the blood concentration at which a drug becomes toxic and the concentration at which the drug is effective. Stated differently, the TI is a comparison of the amount of a therapeutic agent that causes the therapeutic effect to the amount that causes toxicity (Trevor et al., 2013). The larger the TI, the safer the drug is. If the TI is small, as is the case with lithium, then the drug must be administered (dosed) carefully and the person receiving the lithium should be monitored closely for any signs of lithium toxicity. The following is used to determine TI:

Therapeutic Index $= \dfrac{LD_{50}}{ED_{50}}$ in animal studies, or for humans, Therapeutic Index $= \dfrac{TD_{50}}{ED_{50}}$

where

LD_{50} is the lethal dose for 50% of the population.
ED_{50} is the efficacious, effective dose in 50% of subjects.
TD_{50} is the toxic dose in 50% of subjects.

Sometimes the term *safety ratio* is used instead of TI. This is especially case when referring to psychoactive drugs used for non-therapeutic purposes (e.g., recreational use; Gable, 2004). In such cases, the *effective* dose is the amount and frequency that produces the *desired* effect. Drug interactions or synergistic effects are not considered in the TI. Also, the TI does not consider what the ease or difficulty of reaching a toxic or lethal dose. Note that this is more of a consideration for recreational drug use.

A similar concept it called the *protective index*—the difference is that it uses TD_{50} (median toxic dose) in place of LD_{50}. Note that for many substances, toxic effects can occur at levels below those needed to cause death, and because of this the protective index is often more information about a substance's relative safety (i.e., if toxicity is properly specified). Even so the therapeutic index with its advantages of objectivity and easier comprehension is still useful as it can be considered an upper bound for the protective index.

Another term commonly associated or used with therapeutic index is *therapeutic window* which is the range of drug dosages can treat disease effectively without having toxic effects (Ratanajamit et al., 2006). To avoid harm, medication with a small therapeutic window must be administered with care and control. Lithium has a narrow therapeutic window.

Optimal biological dose is the quantity of a drug that will most effectively produce the desired effect while remaining in the range of acceptable toxicity. The purpose of dose finding in disease is to find the maximum tolerated dose based solely on toxicity. Then again, for molecularly targeted agents, little toxicity may arise within the therapeutic dose range and the dose-response curve may not be monotonic. This challenges the principle that more is better.

Pharmacological treatment that will produce the desired effect without unacceptable toxicity is known as the *maximum tolerated dose* (MTD) (National Cancer Institute, 2010; Congressional Research, 2005). MTD studies are done in clinical trials.

So, with regard to lithium, it has a narrow therapeutic index and poisonings are mostly due to chronic poisonings secondary to impaired renal elimination. Renal side effects associated with lithium include polyuria (excessive production and passage of urine), nephrogenic diabetes insipidus (disorder of water balance), proteinuria (presence of protein, usually albumin, in urine), distal renal tubular acidosis (leads to alkaline urine), and reduction in glomerular filtration rate (volume of fluid filtered from the kidney).

DID YOU KNOW?

In the ongoing research in medicine, researchers are continually attempting to find preventors and cures for various diseases. From trying to find the silver bullet for all forms of cancer and amyotrophic lateral sclerosis (ALS), for example, to other deadly diseases and other debilitating maladies such as Alzheimer's disease, the research is non-stop. This research also applies to the use of lithium as a potential cure-all substance. For instance, tentative evidences in Alzheimer's disease show that lithium may actually have slow progression (Forlenza et al., 2012). However, the jury is still out on deciding if lithium can or will cure or prevent ALS—so far, a study has shown that lithium had no effect on ALS outcomes (Ludolph, et al., 2012). The bottom line on the cure and prevention of diseases is that at this stage of research of the efficacy of lithium, we simply do not know what we do not know.

The bottom line is that future studies are needed to discover the exact kidney protective dose and test the effects of low-dose lithium on acute and chronic kidney disease in humans.

HYPERPARATHYROIDISM

Earlier, it was pointed out that hyperparathyroidism (parathyroid glands create high amounts of parathyroid hormone in the bloodstream) and consequent hypercalcemia (calcium level in blood is above normal) can also arise during long-term treatment with lithium. Lithium-associated hyperparathyroidism is the leading cause of hyper-calcemia (calcium level in blood in above normal) or causes an increased set point of calcium for parathyroid hormone suppression and leads to parathyroid hyperplasia (a higher level of calcium in blood).

REFERENCES

Aiff, H., Attman, P.O., Aurel, M., Bendz, H., Ramsauer, B., Schon, S., and Svedlund, J. (2015). Effects of 10 to 30 years of lithium treatment on kidney function. *Journal of Psychopharmacology*, v. 29, no. 5, pp. 608–614.

Australian Medicines Handbook (2013). *Australian Medicine Handbook, Pty. Ltd.* Adelaide: Australia.

Baek, J.H., Kinrys, G., and Nierenbert, A.A. (2013), Lithium tremor revisited: Pathophysiology and treatment. *Acta Psychiatrica Scandinavica*, v. 129, no. 1, pp. 17–23.

Congressional Research Service (2005). *Report for Congress: Agriculture: A glossary of terms, programs and laws, 2005 edition.* Washington, DC: Congressional Research Service.

DruaPoint®System (2013). *Truven Health Analytics, Inc.* Greenwood Village, CO: Thomsen Healthcare.

Duma, S.R., and Fung, V.S. (2019). Drug-induced movement disorders. *Australian Prescriber*, v. 42, no. 2, pp. 56–61.

Fairbrother, F., Petzel, N., Scott, J.G., and Kisley, S. (2019). Lithium can cause hyperthyroid-ism as well as hypothyroidism: A systematic review of an underrecognized association. *Australian and New Zealand Journal of Psychiatry*, v. 53, no. 5, pp. 384–402.

Forlenza, L., et al. (2012). Does lithium prevent Alzheimer's disease. *Drugs Aging*, v. 29, no. 5, p. 338.

Francescangeli, J., Karamchandani, K., Powell, M., and Bonavia, A. (2019). The serotonin syndrome: From molecular mechanisms to clinical practice. *International Journal of Molecular Sciences*, v. 20, no. 9, p. 2288.

Gable, R.S. (2004). Comparison of acute lethal toxicity of commonly abused psychoactive substances. *Addiction*, v. 99, no. 6, pp. 686–696.

Gitlin, M. (2021). Lithium side effects and toxicity prevalence and management strategies. *International Journal of Bipolar Disorders*, v. 4, no. 1, p. 27.

Hilder, J.M., and Rymer, Z.W. (1999). A stimulation study of reflex instability in spastic-ity: Origins of Clonus. *IEEE Transactions on Rehabilitation Engineering*, v. 7, no. 3, pp. 327–340.

Joint Formulary Committee (2013). *British national formulary.* London: British National Formulary, pp. 240–242.

Lithium (Rx) (2013). Eskalith, L. *Medscape, WebMD.* Accessed 12/4/22 @ http://reference.medscape.com/drug/eskalight-lithobid=lithium-342934.

Ludolph, A., Brettschneider, J., and Weishaupt, J.H. (2012). Amyotrophic lateral scleroses. *Current Opinion in Neurology*, v. 25, no. 5, pp. 530–535.

Malhi, G.S., Tanious, M., Bargh, D., Das, P., and Berk, M. (2013). Safe and effective use of lithium. *Australian Prescriber*, v. 36, pp. 18–21.

National Cancer Institute (2010). Dictionary of cancer terms. Accessed 12/02/22 @ http://www/cancer.gov/dictionary/?CdrID=546597.

National library of Medicine (2013). Lithobid tablet, film coated, extended release. Accessed 12/04/22 @ http://dailymed.nim, nih.gov/dailymed/lookup.cf.

Piziak, V.K., Sellman, J.E., and Othmer, E. (1978). Lithium and hypothyroidism. *Journal of Clinical Psychiatry*, v. 39, no. 9, pp. 709–711.

Ratanajamit. C., Soorapan, S., Doang-ngern, T., Waenwaisart, W., Suwanchavalit, L., Suwansiri, S., Jantasaro, S., and Yanatte, I. (2006). Appropriateness of therapeutic drug monitoring for lithium. *Journal of Medical Association of Thailand*, v. 89, no. 11, pp. 1954–1960.

Simon, L.V., and Keenaghan, M (2022). Serotonin syndrome, in StatPearls Publisher. Accessed 11/30/22 @ https://www.ncbi.nlm.nih.gov/books/nbk48237711.

Sperner-Unterweger, B., Fleischhacker, W.W., and Kaschka, W. P. (2001). *Psychoneuroimmunology*. Basil: Karger Publishers.

Srivastava, A. Singh, P., Gupta, H., Kaur, H., Kanojia, N., Guin, D. Sood, M., Chadda, R.K., Yadav, J., Vohora, D., Saso, P., and Kukreti, R. (2019). Systems approach to identify common genes and pathways associated with response to selective serotonin reuptake inhibitors and major depression risk. *International Journal of Molecular sciences*, v. 20, no. 8, p. 1993.

TGA eBusiness Services (2017). *Product information Lithicarb (Lithium carbonate)*. *Psychoneuroimmunology*. Basil: Karger Publishers.

Trevor, A., Katzung, B., Masters, S., and Knuidering-Hall, M. (2013). Chapter 2: Pharmacodynamics, in *Pharmacology examination & board review* (10th ed.). New York: McGraw-Hill Medical, p. 17.

Part 4

Lithium Power Applications

13 Understanding Electricity

INTRODUCTION

Lithium is widely used in non-rechargeable batteries and in rechargeable lithium storage batteries. It is important to point out that batteries do not produce or store electricity; instead they store energy. Energy is of several types:

- kinetic (motion) energy
- water energy
- potential (at rest)
- elastic energy
- nuclear energy
- chemical energy
- sound energy
- internal energy
- heat/thermal energy
- light/radiation energy
- electric energy

In this book, it is electric energy that is our focus, particularly the electrical energy stored in lithium-type batteries.

Note: Based on years of teaching college-level courses, where the subject areas included lithium-based energy applications, I found that in order to better grasp electrical energy stored in lithium batteries that are used in several different applications, it is first important to understand electricity—what is electricity and so forth.

BASIC ELECTRICITY[1]

People living and working in modern societies generally have little difficulty recognizing electrical equipment—electrical equipment is everywhere and (if one pays attention to his or her surroundings) is easy to spot. Despite its great importance in our daily lives, however, few of us probably stop to think what life would be like without electricity. Like air and water, we tend to take electricity for granted. But we use electricity to do many jobs for us every day—from lighting, heating, and cooling our homes to powering our televisions and computers. Then there is the workplace—can we actually perform work without electricity? For example, the typical industrial workplace is outfitted with electrical equipment that

- generates electricity (a generator or emergency generator)
- stores electricity (batteries—batteries store electrical energy and do not create it)
- changes electricity from one level (voltage or current) to another (transformers)
- transports or transmits and distributes electricity throughout the plant site (wiring distribution systems)
- measures electricity (measuring meters/indicators)
- converts electricity into other forms of energy (rotating shafts—mechanical energy, heat energy, light energy, chemical energy, or radio energy)
- protects other electrical equipment (fuses, circuit breakers, or relays)
- operates and controls other electrical equipment (motor controllers)
- converts some condition or occurrence into an electric signal (sensors)
- converts some measured variable to a representative electrical signal (transducers or transmitters)

NATURE OF ELECTRICITY

The word *electricity* is derived from the Greek word "electron" (meaning AMBER). Amber is a translucent (semitransparent) yellowish fossilized mineral resin. The ancient Greeks used the words "electric force" in referring to the mysterious forces of attraction and repulsion exhibited by amber when it was rubbed with a cloth. They did not understand the nature of this force. They could not answer the question, "What is electricity?" This question still remains unanswered. Today, we often attempt to answer this question by describing the effect and not the force. That is, the standard answer given in physics is: Electricity is "the force that moves electrons," which is about the same as defining a sail as "that force which moves a sailboat."

At the present time, little more is known than the ancient Greeks knew about the fundamental nature of electricity, but we have made tremendous strides in harnessing and using it. As with many other unknown (or unexplainable) phenomena, elaborate theories concerning the nature and behavior of electricity have been advanced and have gained wide acceptance because of their apparent truth—and because they work.

Scientists have determined that electricity seems to behave in a constant and predictable manner in given situations or when subjected to given conditions. Faraday, Ohm, Lenz, and Kirchhoff have described the predictable characteristics of electricity and electric current in the form of certain rules. These rules are often referred to as laws. Thus, though electricity itself has never been clearly defined, its predictable nature and easily used energy form have made it one of the most widely used power sources in modern times.

The bottom line: You can "learn" about electricity by learning the rules, or laws, applying to the behavior of electricity and by understanding the methods of producing, controlling, and using it. Thus, this learning about electricity can be accomplished without ever having determined its fundamental identity.

You are probably scratching your head, puzzled.

We understand the main question running through your brain at this exact moment: "This is a section in the text about the physics of electricity and the authors can't even explain what electricity is?"

That is correct; we cannot. The point is no one can definitively define electricity. Electricity is one of those subject areas where the old saying "we don't know what we don't know about it" fits perfectly.

Again, there are a few theories about electricity that have so far stood the test of extensive analysis and much time (relatively speaking, of course). One of the oldest and the most generally accepted theories, concerning electric current flow (or electricity), is known as the *electron theory*.

The electron theory basically states that electricity or current flow is the result of the flow of free electrons in a conductor. Thus, electricity is the flow of free electrons or simply electron flow. And, in this text, this is how we define electricity; that is, again, electricity is the flow of free electrons.

Electrons are extremely tiny particles of matter. To gain understanding of electrons and exactly what is meant by "electron flow," it is necessary to briefly discuss the structure of matter.

Atoms are composed, in various combinations, of subatomic particles of *electrons, protons*, and *neutrons*. These particles differ in weight (a proton is much heavier than the electron) and charge. We are not concerned with the weights of particles in this text, but the *charge* is extremely important in electricity. The electron is the fundamental negative charge (–) of electricity. Electrons revolve about the nucleus or center of the atom in paths of concentric *orbits*, or shells. The proton is the fundamental positive (+) charge of electricity. Protons are found in the nucleus. The number of protons within the nucleus of any particular atom specifies the atomic number of that atom. For example, the helium atom has two protons in its nucleus, so the atomic number is 2. The neutron, which is the fundamental neutral charge of electricity, is also found in the nucleus.

Most of the weight of the atom is in the protons and neutrons of the nucleus. Whirling around the nucleus is one or more negatively charged electrons. Normally, there is one proton for each electron in the entire atom so that the net positive charge of the nucleus is balanced by the net negative charge of the electrons rotating around the nucleus (see Chapter 1, Figure 1.3).

Important point: Most batteries are marked with the symbols + and – or even with the abbreviations POS (positive) and NEG (negative). The concept of a positive or negative polarity and its importance in electricity will become clear later. However, for the moment, you need to remember that an electron has a negative charge and that a proton has a positive charge.

We stated earlier that in an atom the number of protons is usually the same as the number of electrons. This is an important point because this relationship determines the kind of element (the atom is the smallest particle that makes up an element; an element retains its characteristics when subdivided into atoms) in question. Recall that Figure 1.4 in Chapter 1 shows a simplified drawing of several atoms of different materials based on the conception of electrons orbiting about the nucleus. For example, hydrogen has a nucleus consisting of one proton, around which rotates one electron. The helium atom has a nucleus containing two protons and two neutrons with two electrons encircling the nucleus. Both of these elements are electrically neutral (or balanced) because each has an equal number of electrons and protons. Since the negative (–) charge of each electron

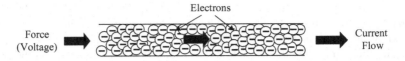

FIGURE 13.1 Electron flow in a copper wire.

is equal in magnitude to the positive (+) charge of each proton, the two opposite charges cancel.

A balanced (neutral or stable) atom has a certain amount of energy, which is equal to the sum of the energies of its electrons. Electrons, in turn, have different energies called *energy levels*. The energy level of an electron is proportional to its distance from the nucleus. Therefore, the energy levels of electrons in shells farther from the nucleus are higher than that of electrons in shells nearer the nucleus.

When an electric force is applied to a conducting medium, such as copper wire, electrons in the outer orbits of the copper atoms are forced out of orbit (i.e., liberating or freeing electrons) and impelled along the wire. This electrical force, which forces electrons out of orbit, can be produced in a number of ways, such as: by moving a conductor through a magnetic field; by friction, as when a glass rod is rubbed with cloth (silk); or by chemical action, as in a battery.

When the electrons are forced from their orbits, they are called *free electrons*. Some of the electrons of certain metallic atoms are so loosely bound to the nucleus that they are relatively free to move from atom to atom. These free electrons constitute the flow of an electric current in electrical conductors.

Important point: When an electric force is applied to a copper wire, free electrons are displaced from the copper atoms and move along the wire, producing electric current as shown in Figure 13.1.

If the internal energy of an atom is raised above its normal state, the atom is said to be *excited*. Excitation may be produced by causing the atoms to collide with particles that are impelled by an electric force. In effect what occurs is that energy is transferred from the electric source to the atom. The excess energy absorbed by an atom may become sufficient to cause loosely bound outer electrons to leave the atom against the force that acts to hold them within.

Important point: An atom that has lost or gained one or more electrons is said to be *ionized*. If the atom loses electrons, it becomes positively charged and is referred to as a *positive ion*. Conversely, if the atom gains electrons, it becomes negatively charged and is referred to as a *negative ion*.

CONDUCTORS

Recall that we pointed out earlier that electric current moves easily through some materials but with greater difficulty through others. Three good electrical conductors are copper, silver, and aluminum (generally, we can say that most metals are good conductors.) At the present time, copper is the material of choice used in electrical conductors. Under special conditions, certain gases are also used as conductors (e.g., neon gas, mercury vapor, and sodium vapor are used in various kinds of lamps).

The function of the wire conductor is to connect a source of applied voltage to a load resistance with a minimum IR voltage drop in the conductor so that most of the applied voltage can produce current in the load resistance. Ideally, a conductor must have a very low resistance (e.g., a typical value for a conductor—copper—is less than $1\Omega/10$ ft).

Because all electrical circuits utilize conductors of one type or another, in this section we discuss the basic features and electrical characteristics of the most common types of conductors.

Moreover, because conductor splices and connections (and insulation of such connections) are also an essential part of any electric circuit, they are also discussed.

UNIT SIZE OF CONDUCTORS

A standard (or unit size) of a conductor has been established to compare the resistance and size of one conductor with another. The unit of linear measurement used is (in regard to diameter of a piece of wire) the **mil** (0.001 of an inch). A convenient unit of wire length used is the **foot**. Thus, the standard unit of size in most cases is the **mil-foot** (i.e., a wire will have unit size if it has diameter of 1 mil and a length of 1 ft). The resistance in ohms of a unit conductor or a given substance is called the **resistivity** (or specific resistance) of the substance.

As a further convenience, **gage** numbers are also used in comparing the diameter of wires. The B and S (Browne and Sharpe) gage was used in the past; now the most commonly used gage is the **American Wire Gage** (AWG).

SQUARE MIL

Figure 13.2 shows a square mil. The *square mil* is a convenient unit of cross-sectional area for square or rectangular conductors. As shown in Figure 13.2, a square mil is the area of a square, the sides of which are 1 mil. To obtain the cross-sectional area in square mil of a square conductor, square one side measured in mil. To obtain the cross-sectional area in square mil of a rectangular conductor, multiply the length of one side by that of the other, each length being expressed in mil.

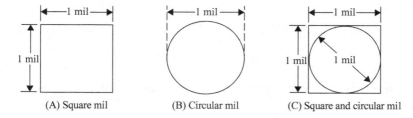

(A) Square mil   (B) Circular mil   (C) Square and circular mil

FIGURE 13.2   (a) Square mil; (b) circular mil; and (c) comparison of circular to square mil.

THE 411 ON SQUARE MIL

A square mil is a unit of area, equal to the area of a square with sides of length one mil, and is often used in specifying the area of the cross section of a wire or cable.

1 square mil is equal to:

- 1 millionth of a square inch (1 in^2 is equal to 1 million square mil)
- 6.4516×10^{-10} m^2
- about 1.273 circular mil

Example 13.1

Problem:

Find the cross-sectional area of a large rectangular conductor 5/8 in thick and 5 in wide.

Solution:

The thickness may be expressed in mil as $0.625 \times 1,000 = 625$ mil and the width as $5 \times 1,000 = 5,000$ mil. The cross-sectional area is $625 \times 5,000$, or 3,125,000 square mil.

CIRCULAR MIL

The *circular mil* is the standard unit of wire cross-sectional area used in most wire tables. To avoid the use of decimals (because most wires used to conduct electricity may be only a small fraction of an inch), it is convenient to express these diameters in mil. For example, the diameter of a wire is expressed as 25 mil instead of 0.025 in. A circular mil is the area of a circle having a diameter of 1 mil, as shown in Figure 13.2b. The area in circular mil of a round conductor is obtained by squaring the diameter measured in mil. Thus, a wire having a diameter of 25 has an area of 25^2 or 625 circular mil. By way of comparison, the basic formula for the area of a circle is

$$A = \pi R^2 \tag{13.1}$$

and in this example the area in square inches is

$$A = \pi R^2 = 3.14(0.0125)^2 = 0.00049 \text{ in}^2$$

If D is the diameter of a wire in mil, the area in square-mil area can be determined using

$$A = \pi(D/2)^2 \tag{13.2}$$

which translates to

$$= 3.14 / 4 \; D^2$$

$$= 0.785 \; D^2 \text{sq. mil}$$

Thus, a wire 1 mil in diameter has an area of

$$A = 0.785 \times 1^2 = 0.785 \text{ sq. mil,}$$

which is equivalent to 1 circular mil. The cross-sectional area of a wire in circular mil is therefore determined as

$$A = \frac{0.785 \; D^2}{0.785} = D^2 \text{ circular mil,}$$

where D is the diameter in mil. Therefore, the constant $\pi/4$ is eliminated from the calculation.

Note that in comparing square and round conductors that the circular mil is a smaller unit of area than the square mil, and therefore there are more circular mil than square mil in any given area. The comparison is shown in Figure 13.2c. The area of a circular mil is equal to 0.785 of a square mil.

Important point: To determine the circular-mil area when the square-mil area is given, divide the area in square mil by 0.785. Conversely, to determine the square-mil area when the circular-mil area is given, multiply the area in circular mil by 0.785.

Example 13.2

Problem:

A No. 12 wire has a diameter of 80.81 mil. What is (1) its area in circular mil and (2) its area in square mil?

Solution:

1. $A = D^2 = 80.81^2 = 6{,}530$ circular mil
2. $A = 0.785 \times 6{,}530 = 5{,}126$ square mil

Example 13.3

Problem:

A rectangular conductor is 1.5 in wide and 0.25 in thick. (1) What is its area in square mil? (2) What size of round conductor in circular mil is necessary to carry the same current as the rectangular bar?

Solution:

1. 1.5 in = 1.5 × 1,000 = 1,500 mil
   0.25 in = 0.25 × 1,000 = 250 mil
   A = 1,500 × 250 = 375,000 sq. mil
2. To carry the same current, the cross-sectional area of the rectangular bar and the cross-sectional area of the round conductor must be equal. There are more circular mil than square mil in this area, and therefore

$$A = \frac{375,000}{0.785} = 477,700 \text{ circular mil}$$

Note: Many electric cables are composed of stranded wires. The strands are usually single wires twisted together in sufficient numbers to make up the necessary cross-sectional area of the cable. The total area in circular mil is determined by multiplying the area of one strand in circular mil by the number of strands in the cable.

CIRCULAR-MIL FOOT

As shown in Figure 13.3, a *circular-mil-foot* is actually a unit of volume. More specifically, it is a unit conductor 1 foot in length and having a cross-sectional area of 1 circular mil. The circular-mil-foot is useful in making comparisons between wires that are made of different metals because it is considered a unit conductor. Because it is considered a unit conductor, the circular-mil-foot is useful in making comparisons between wires that are made of different metals. For example, a basis of comparison of the **resistivity** of various substances may be made by determining the resistance of a circular-mil-foot of each of the substances.

Note: It is sometimes more convenient to employ a different unit of volume when working with certain substances. Accordingly, the unit Volume may also be taken as the centimeter cube. The inch cube may also be used. The unit of volume employed is given in tables of specific resistances.

RESISTIVITY

All materials differ in their atomic structure and therefore in their ability to resist the flow of an electric current. The measure of the ability of a specific material to resist the flow of electricity is called its *resistivity*, or specific resistance—the resistance in ohms offered by unit volume (the circular-mil-foot) of a substance to the flow of

FIGURE 13.3   Circular-mil-foot.

electric current. Resistivity is the reciprocal of conductivity (i.e., the ease by which current flows in a conductor). A substance that has a high resistivity will have a low conductivity, and vice versa.

The resistance of a given length, for any conductor, depends upon the resistivity of the material, the length of the wire, and the cross-sectional area of the wire according to the equation

$$R = \rho \frac{L}{A} \qquad (13.3)$$

where

R = resistance of the conductor, Ω
L = length of the wire, ft
A = cross-sectional area of the wire, CM
ρ = specific resistance or resistivity, CM × Ω/ft

The factor ρ (Greek letter rho, pronounced "roe") permits different materials to be compared for resistance according to their nature without regard to different lengths or areas. Higher values of ρ mean more resistance.

Key point: The resistivity of a substance is the resistance of a unit volume of that substance.

Many tables of resistivity are based on the resistance in ohms of a volume of the substance 1 ft long and 1 circular mil in cross-sectional area. The temperature at which the resistance measurement is made is also specified. If the kind of metal of which the conductor is made is known, the resistivity of the metal may be obtained from a table. The resistivity, or specific resistance, of some common substances are given in Table 13.1.

Note: Since silver, copper, gold, and aluminum have the lowest values of resistivity, they are the best conductors. Tungsten and iron have a much higher resistivity.

TABLE 13.1
Resistivity (Specific Resistance)

Substance	Specific Resistance @ 20° (CM ft Ω)
Silver	9.8
Copper (drawn)	10.37
Gold	14.7
Aluminum	17.02
Tungsten	33.2
Brass	42.1
Steel (soft)	95.8
Nichrome	660.0

Example 13.4

Problem:

What is the resistance of 1,000 ft of copper wire having a cross-sectional area of 10,400 circular mil (No. 10 wire), the wire temperature being 20°C?

Solution:

The resistivity (specific resistance), from Table 13.1, is 10.37. Substituting the known values in the preceding equation (13.3), the resistance, R, is determined as

$$R = \rho \frac{L}{A} = 10.37 \times \frac{1,000}{10,400} = 1\Omega, \text{ approximately}$$

WIRE MEASUREMENT

Wires are manufactured in sizes numbered according to a table known as the AWG. Table 13.2 lists the standard wire sizes that correspond to the AWG. The gage numbers specify the size of round wire in terms of its diameter and cross-sectional area. Note the following:

   a. As the gage numbers increase from 1 to 40, the diameter and circular area decrease. Higher gage numbers mean smaller wire sizes. Thus, No. 12 is a smaller wire than No. 4.
   b. The circular area doubles for every three gage sizes. For example, No 12 wire has about twice the area of No. 15 wire.
   c. The higher the gage number and the smaller the wire, the greater the resistance of the wire for any given length. Therefore, for 1,000 ft of wire, No. 12 has a resistance of 1.62Ω while No. 4 has 0.253Ω.

FACTORS GOVERNING SELECTION OF WIRE SIZE

Several factors must be considered in selecting the size of wire to be used for transmitting and distributing electric power. These factors include allowable power loss in the line; the permissible voltage drop in the line; the current-carrying capacity of the line; and the ambient temperatures in which the wire is to be used.

   a. *Allowable power loss (I^2R) in the line.* This loss represents electrical energy converted into heat. The use of large conductors will reduce the resistance and therefore the I^2R loss. However, large conductors are heavier and require more substantial supports; thus, they are more expensive initially than small ones.
   b. *Permissible voltage drop (IR drop) in the line.* If the source maintains a constant voltage at the input to the line, any variation in the load on the line will cause a variation in line current, and a consequent variation in the IR drop

TABLE 13.2
Copper wire table

Gage #	Diameter	Circular mil	Ohms/1,000 ft @ 25°C
1	289.0	83,700.0	0.126
2	258.0	66,400.0	0.159
3	229.0	52,600.0	0.201
4	204.0	41,700.0	0.253
5	182.0	33,100.0	0.319
6	162.0	26,300.0	0.403
7	144.0	20,800.0	0.508
8	128.0	16,500.0	0.641
9	114.0	13,100.0	0.808
10	102.0	10,400.0	1.02
11	91.0	8,230.0	1.28
12	81.0	6,530.0	1.62
13	72.0	5,180.0	2.04
14	64.0	4,110.0	2.58
15	57.0	3,260.0	3.25
16	51.0	2,580.0	4.09
17	45.0	2,050.0	5.16
18	40.0	1,620.0	6.51
19	36.0	1,290.0	8.21
20	32.0	1,020.0	10.4
21	28.5	810.0	13.1
22	25.3	642.0	16.5
23	22.6	509.0	20.8
24	20.1	404.0	26.4
25	17.9	320.0	33.0
26	15.9	254.0	41.6
27	14.2	202.0	52.5
28	12.6	160.0	66.2
29	11.3	127.0	83.4
30	10.0	101.0	105.0
31	8.9	79.7	133.0
32	8.0	63.2	167.0
33	7.1	50.1	211.0
34	6.3	39.8	266.0
35	5.6	31.5	335.0
36	5.0	25.0	423.0
37	4.5	19.8	533.0
38	4.0	15.7	673.0
39	3.5	12.5	848.0
40	3.1	9.9	1,070.0

in the line. A wide variation in the IR drop in the line causes poor voltage regulation at the load.

c. *The current-carrying capacity of the line.* When current is draw through the line, heat is generated. The temperature of the line will rise until the heat radiated, or otherwise dissipated, is equal to the heat generated by the passage of current through the line. If the conductor is insulated, the heat generated in the conductor is not so readily removed as it would be if the conductor were not insulated.

d. *Conductors installed where ambient temperature is relatively high.* When installed in such surroundings, the heat generated by external sources constitutes an appreciable part of the total conductor heating. Due allowance must be made for the influence of external heating on the allowable conductor current and each case has its own specific limitations.

COPPER VS OTHER METAL CONDUCTORS

If it were not cost prohibitive, silver, the best conductor of electron flow (electricity), would be the conductor of choice in electrical systems. Instead, silver is used only in special circuits where a substance with high conductivity is required.

The two most generally used conductors are copper and aluminum. Each has characteristics that make its use advantageous under certain circumstances. Likewise, each has certain disadvantages, or limitations.

In regard to **copper**, it has a higher conductivity; it is more ductile (can be drawn out into wire), has relatively high tensile strength, and can be easily soldered. It is more expensive and heavier than aluminum.

Aluminum has only about 60% of the conductivity of copper, but its lightness makes possible long spans, and its relatively large diameter for a given conductivity reduces corona (i.e., the discharge of electricity from the wire when it has a high potential). The discharge is greater when smaller diameter wire is used than when larger diameter wire is used. However, aluminum conductors are not easily soldered, and aluminum's relatively large size for a given conductance does not permit the economical use of an insulation covering.

Note: Recent practice involves using copper wiring (instead of aluminum wiring) in house and some industrial applications. This is the case because aluminum connections are not as easily made as they are with copper. In addition, over the years, many fires have been started because of improperly connected aluminum wiring (i.e., poor connections = high resistance connections, resulting in excessive heat generation).

A comparison of some of the characteristics of copper and aluminum is given in Table 13.3.

TEMPERATURE COEFFICIENT (WIRE)

The resistance of pure metals—such as silver, copper, and aluminum—increases as the temperature increases. The *temperature coefficient* of resistance, α (Greek letter alpha), indicates how much the resistance changes for a change in temperature.

TABLE 13.3
Characteristics of Copper & Aluminum

Characteristics	Copper	Aluminum
Tensile strength (lb/in²)	55,000	25,000
Tensile strength for same conductivity (lb)	55,000	40,000
Weight for same conductivity (lb)	100	48
Cross-section for same conductivity (CM)	100	160
Specific resistance (Ω/mil. ft)	10.6	17

TABLE 13.4
Properties of Conducting Materials (Approximate)

Material	Temperature Coefficient, Ω/°C
Aluminum	0.004
Carbon	−0.0003
Constantan	0 (average)
Copper	0.004
Gold	0.004
Iron	0.006
Nichrome	0.0002
Nickel	0.005
Silver	0.004
Tungsten	0.005

A positive value for α means R increases with temperature; a negative α means R decreases; and a zero α means R is constant, not varying with changes in temperature. Typical values of α are listed in Table 13.4.

The amount of increase in the resistance of a 1-Ω sample of the copper conductor per degree rise in temperature (i.e., the temperature coefficient of resistance) is approximately 0.004. For pure metals, the temperature coefficient of resistance ranges between 0.004 and 0.006Ω.

Thus, a copper wire having a resistance of 50Ω at an initial temperature of 0°C will have an increase in resistance of 50 × 0.004, or 0.2Ω (approximate) for the entire length of wire for each degree of temperature rise above 0°C. At 20°C, the increase in resistance is approximately 20 × 0.2, or 4Ω. The total resistance at 20°C is 50 + 4, or 54Ω.

Note: As shown in Table 13.4, carbon has a negative temperature coefficient. In general, α is negative for all semiconductors such as germanium and silicon. A negative value for α means less resistance at higher temperatures. Therefore, the resistance of semiconductor diodes and transistors can be reduced considerably when they become hot with normal load current. Observe, also, that constantan has a value of zero for α (Table 13.4). Thus, it can be used for precision wire-wound resistors, which do not change resistance when the temperature increases.

CONDUCTOR INSULATION

Electric current must be contained; it must be channeled from the power source to a useful load – safely. To accomplish this, electric current must be forced to flow only where it is needed. Moreover, current-carrying conductors must not be allowed (generally) to come in contact with one another, their supporting hardware, or personnel working near them. To accomplish this, conductors are coated or wrapped with various materials. These materials have such a high resistance that they are, for all practical purposes, nonconductors. They are generally referred to as *insulators* or *insulating materials.*

There are a wide variety of insulated conductors available to meet the requirements of any job. However, only the necessary minimum of insulation is applied for any particular type of cable designed to do a specific job. This is the case because insulation is expensive and has a stiffening effect and is required to meet a great variety of physical and electrical conditions.

Two fundamental, but distinctly different properties of insulation materials (e.g., rubber, glass, asbestos, and plastics) are insulation resistance and dielectric strength.

   a. *Insulation resistance* is the resistance to current leakage through and over the surface of insulation materials.
   b. *Dielectric strength* is the ability of the insulator to withstand potential difference and is usually expressed in terms of the voltage at which the insulation fails because of the electrostatic stress.

Various types of materials are used to provide insulation for electric conductors, including rubber, plastics, varnished cloth, paper, silk, cotton, and enamel.

CONDUCTOR SPLICES & TERMINAL CONNECTIONS

When conductors join each other, or connect to a load, *splices* or *terminals* must be used. It is important that they be properly made, since any electric circuit is only as good as its weakest connection. The basic requirement of any splice or connection is that it be both mechanically and electrically as strong as the conductor or device with which it is used. High-quality workmanship and materials must be employed to ensure lasting electrical contact, physical strength, and insulation (if required).

Important point: Conductor splices and connections are an essential part of any electric circuit.

SOLDERING OPERATIONS

Soldering operations are a vital part of electrical and/or electronics maintenance procedures. Soldering is a manual skill that must be learned by all personnel who work in the field of electricity. Obviously, practice is required to develop proficiency in the techniques of soldering.

In performing a soldering operation both the solder and the material to be soldered (e.g., electric wire and/or terminal lugs) must be heated to a temperature which allows the solder to flow. If either is heated inadequately, **cold** solder joints result (i.e., high resistance connections are created). Such joints do not provide either the physical strength or the electrical conductivity required. Moreover, in soldering operations, it is necessary to select a solder that will flow at a temperature low enough to avoid damage to the part being soldered, or to any other part or material in the immediate vicinity.

SOLDERLESS CONNECTORS

Generally, terminal lugs and splicers which do not require solder are more widely used (because they are easier to mount correctly) than those which do require solder. Solderless connectors—made in a wide variety of sizes and shapes—are attached to their conductors by means of several different devices, but the principle of each is essentially the same. They are all crimped (squeezed) tightly onto their conductors. They afford adequate electrical contact, plus great mechanical strength.

INSULATION TAPE

The carpenter has his saw, the dentist his pliers, the plumber his wrench, and the electrician his insulating tape. Accordingly, one of the first things the rookie maintenance operator learns (a rookie who is also learning proper and safe techniques for performing electrical work) is the value of electrical insulation tape. Normally, the use of electrical insulating tape comes into play as the final step in completing a splice or joint, to place insulation over the bare wire at the connection point.

Typically, insulation tape used should be the same basic substance as the original insulation, usually a rubber-splicing compound. When using rubber (latex) tape as the splicing compound where the original insulation was rubber, it should be applied to the splice with a light tension so that each layer presses tightly against the one underneath it. In addition to the rubber tape application (which restores the insulation to original form), restoring with friction tape is also often necessary.

In recent years, plastic electrical tape has come into wide use. It has certain advantages over rubber and friction tape. For example, it will withstand higher voltages for a given thickness. Single thin layers of certain commercially available plastic tape will stand several thousand volts without breaking down.

Important point: Be advised that though the use of plastic electrical tape has become almost universal in industrial applications, it must be applied in more layers—because it is thinner than rubber or friction tape—to ensure an extra margin of safety.

STATIC ELECTRICITY

Electricity at rest is often referred to as *static electricity*. More specifically, when two bodies of matter have unequal charges, and are near one another, an electric force is exerted between them because of their unequal charges. However, since they are not in contact, their charges cannot equalize. The existence of such an electric force, where current can't flow, is *static electricity.*

However, static, or electricity at rest, will flow if given the opportunity—this is the case because static electricity is an imbalance of negative and positive charges. An example of this phenomenon is often experienced when one walks across a dry carpet and then touches a doorknob—a slight shock is usually felt and a spark at the fingertips is likely noticed. Another familiar example is "static cling." For example, whenever we rub an air-filled balloon against the hair on our heads, and then place the balloon against a wall the balloon sticks to the wall, defying gravity, because of static cling. In the workplace, static electricity is prevented from building up by properly bonding or grounding of equipment to ground or Earth.

CHARGED BODIES

The fundamental law of charged bodies states: "Like charges repel each other and unlike charges attract each other." A positive charge and negative charge, being opposite or unlike, tend to move toward each other—attracting each other. In contrast, like bodies tend to repel each other. Electrons repel each other because of their like negative charges, and protons repel each other because of their like positive charges. Figure 13.4 demonstrates the law of charged bodies.

It is important to point out another significant part of the fundamental law of charged bodies—that is, *force of attraction or repulsion existing between two magnetic poles decreases rapidly as the poles are separated from each other.* More specifically, the force of attraction or repulsion varies directly as the product of the separate pole strengths and inversely as the square of the distance separating the magnetic poles, provided the poles are small enough to be considered as points.

Let's look at an example:

If you increase the distance between two north poles of magnets from 2 feet to 4 feet, the force of repulsion between them is decreased to one fourth of its original value. If either pole strength is doubled, the distance remaining the same, the force between the poles will be doubled.

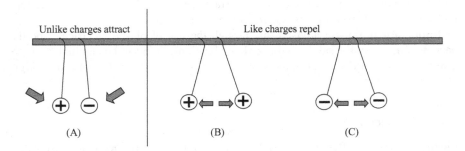

FIGURE 13.4   Reaction between two charged bodies. The opposite charge in (a) attracts. The like charges in (b) and (c) repel each other.

Coulomb's Law

Simply put, Coulomb's Law states that the amount of attracting or repelling force which acts between two electrically charged bodies in free space depends on two things:

a. Their charges
b. The distance between them.

Specifically, *Coulomb's Law* states: "Charged bodies attract or repel each other with a force that is directly proportional to the product of their charges and is inversely proportional to the square of the distance between them."

Note: The magnitude of electric charge a body possesses is determined by the number of electrons compared with the number of protons within the body. The symbol for the magnitude of electric charge is Q, expressed in units of *coulombs* (C). A charge of one positive coulomb means a body contains a charge of 6.25×10^{18}. A charge of one negative coulomb, $-Q$, means a body contains a charge of 6.25×10^{18} more electrons than protons.

Electrostatic Fields

The fundamental characteristic of an electric charge is its ability to exert force. The space between and around charged bodies in which their influence is felt is called an *electric field of force*. The electric field is always terminated on material objects and extends between positive and negative charges. This region of force can consist of air, glass, paper, or a vacuum. This region of force is referred to as an *electrostatic field*.

When two objects of opposite polarity are brought near each other, the electrostatic field is concentrated in the area between them. The field is generally represented by lines that are referred to as *electrostatic lines of force*. These lines are imaginary and are used merely to represent the direction and strength of the field. To avoid confusion, the positive lines of force are always shown leaving charge, and for a negative charge they are shown as entering. Figure 13.5 illustrates the use of lines to represent the field about charged bodies.

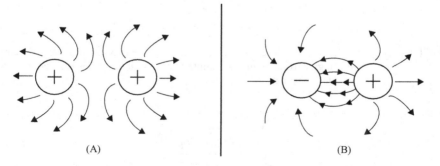

(A)   (B)

FIGURE 13.5   Electrostatic lines of force. (a) Represents the repulsion of like-charged bodies and their associated fields. (b) Represents the attraction between unlike-charged bodies and their associated fields.

Note: A charged object will retain its charge temporarily if there is no immediate transfer of electrons to or from it. In this condition, the charge is said to be *at rest*. Remember, electricity at rest is called *static* electricity.

MAGNETISM

Most electrical equipment depends directly or indirectly upon magnetism. Magnetism is defined as a phenomenon associated with magnetic fields; that is, it has the power to attract such substances as iron, steel, nickel, or cobalt (metals that are known as magnetic materials). Correspondingly, a substance is said to be a magnet if it has the property of magnetism. For example, a piece of iron can be magnetized and is thus a magnet.

When magnetized, the piece of iron (note: we will assume a piece of flat bar 6 inches long x 1-inch-wide x 0.5 inches thick; a bar magnet—see Figure 13.6) will have two points opposite each other which most readily attract other pieces of iron. The points of maximum attraction (one on each end) are called the *magnetic poles* of the magnet: the north (N) pole and the south (S) pole. Just as like electric charges repel each other and opposite charges attract each other, like magnetic poles repel each other and unlike poles attract each other. Although invisible to the naked eye, its force can be shown to exist by sprinkling small iron filings on a glass covering a bar magnet as shown in Figure 13.6.

Figure 13.7 shows how the field looks without iron filings; it is shown as lines of force [known as *magnetic flux or flux lines*; the symbol for magnetic flux is the Greek lowercase letter ϕ (phi)] in the field, repelled away from the north pole of the magnet and attracted to its south pole.

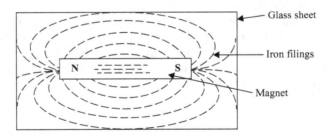

FIGURE 13.6   The magnetic field around a bar magnet. If the glass sheet is tapped gently, the filings will move into a definite pattern that describes the field of force around the magnet.

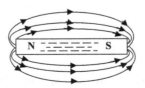

FIGURE 13.7   Magnetic field of force around a bar magnet, indicated by lines of force.

Note: A *magnetic circuit* is a complete path through which magnetic lines of force may be established under the influence of a magnetizing force. Most magnetic circuits are composed largely of magnetic materials in order to contain the magnetic flux. These circuits are similar to the *electric circuit* (an important point), which is a complete path through which current is caused to flow under the influence of an electromotive force.

There are three types or groups of magnets:

a. *Natural magnets*: Found in the natural state in the form of a mineral (an iron compound) called magnetite.
b. *Permanent magnets*: (Artificial magnet), hardened steel or some alloy such as Alinco bars that have been permanently magnetized. The permanent magnet most people are familiar with is the horseshoe magnet; this red U-shaped magnet is the universal symbol of magnets, recognized throughout the world (see Figure 13.8b).

(A)

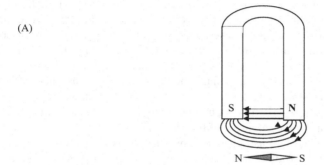

(B)

FIGURE 13.8   (a and b) Horseshoe magnet.

c. *Electromagnets*. (Artificial magnet), composed of soft-iron cores around which are wound coils of insulated wire. When an electric current flows through the coil, the core becomes magnetized. When the current ceases to flow, the core loses most of the magnetism.

Note: Permanent magnets are those of hard magnetic materials (hard steel or alloys) that retain their magnetism when the magnetizing field is removed. A temporary magnet is one that has no ability to retain a magnetized state when the magnetizing field is removed.

Magnetic Materials

Natural magnets are no longer used (they have no practical value) in electrical circuitry because more powerful and more conveniently shaped permanent magnets can be produced artificially. Commercial magnets are made from special steels and alloys—magnetic materials.

Magnetic materials are those materials that are attracted or repelled by a magnet and that can be magnetized themselves. Iron, steel, and alloy bar are the most common magnetic materials. These materials can be magnetized by inserting the material (in bar form) into a coil of insulated wire and passing a heavy direct current through the coil. The same material may also be magnetized if it is stroked with a bar magnet. It will then have the same magnetic property that the magnet used to induce the magnetism has—namely, there will be two poles of attraction, one at either end. This process produces a permanent magnet by induction—that is, the magnetism is induced in the bar by the influence of the stroking magnet.

Even though classified as permanent magnets, it is important to point out that hardened steel and certain alloys are relatively difficult to magnetize and are said to have a *low permeability* because the magnetic lines of force do not easily permeate or distribute themselves readily through the steel.

Note: *Permeability* refers to the ability of a magnetic material to concentrate magnetic flux. Any material that is easily magnetized has high permeability. A measure of permeability for different materials in comparison with air or vacuum is called *relative* permeability, symbolized by μ or (mu).

Once hard steel and other alloys are magnetized; however, they retain a large part of their magnetic strength and are called *permanent magnets*. Conversely, materials that are relatively easy to magnetize—such as soft iron and annealed silicon steel—are said to have a *high permeability*. Such materials retain only a small part of their magnetism after the magnetizing force is removed and are called *temporary magnets*.

The magnetism that remains in a temporary magnet after the magnetizing force is removed is called *residual magnetism*.

Early magnetic studies classified magnetic materials merely as being magnetic and nonmagnetic—that is, based on the strong magnetic properties of iron. However, since weak magnetic materials can be important in some applications, present studies classify materials into one of three groups: namely, paramagnetic, diamagnetic, and ferromagnetic.

a. *Paramagnetic materials.* These include aluminum, platinum, manganese, and chromium—materials that become only slightly magnetized even though under the influence of a strong magnetic field. This slight magnetization is in the same direction as the magnetizing field. Relative permeability is slightly more than 1 (i.e., considered nonmagnetic materials).

b. *Diamagnetic materials.* These include bismuth, antimony, copper, zinc, mercury, gold, and silver—materials that can also be slightly magnetized when under the influence of a very strong field. Relative permeability is less than 1 (i.e., considered nonmagnetic materials).

c. *Ferromagnetic materials.* These include iron, steel, nickel, cobalt, and commercial alloys—materials that are the most important group for applications of electricity and electronics. Ferromagnetic materials are easy to magnetize and have high permeability, ranging 50–3,000.

MAGNETIC EARTH

The Earth is a huge magnet; and surrounding Earth is the magnetic field produced by the Earth's magnetism. Most people would have no problem understanding or at least accepting this statement. However, if told that the Earth's north magnetic pole is actually its south magnetic pole and that the south magnetic pole is actually the Earth's north magnetic pole, they might not accept or understand this statement. But, in terms of a magnet, it is true.

As can be seen from Figure 13.9, the magnetic polarities of the Earth. As shown in Figure 13.9, the geographic poles are also shown at each end of the axis of rotation

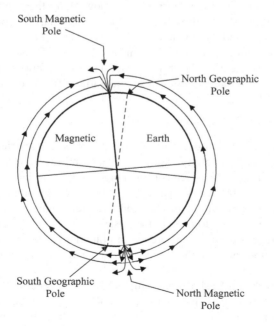

FIGURE 13.9 Earth's magnetic poles.

of the Earth. Clearly, as shown in Figure 13.9, the magnetic axis does not coincide with the geographic axis, and therefore the magnetic and geographic poles are not at the same place on the surface of the Earth.

Recall that magnetic lines of force are assumed to emanate from the north pole of a magnet and to enter the South Pole as closed loops. Because the Earth is a magnet, lines of force emanate from its north magnetic pole and enter the south magnetic pole as closed loops. A compass needle aligns itself in such a way that the Earth's lines of force enter at its south pole and leave at its north pole. Because the north pole of the needle is defined as the end that points in a northerly direction it follows that the magnetic pole in the vicinity of the north geographic pole is in reality a south magnetic pole, vice versa.

DIFFERENCE IN POTENTIAL

Because of the force of its electrostatic field, an electric charge has the ability to do the work of moving another charge by attraction or repulsion. The force that causes free electrons to move in a conductor as an electric current may be referred to as follows:

- Electromotive force (emf)
- Voltage
- Difference in potential

When a difference in potential exists between two charged bodies that are connected by a wire (conductor), electrons (current) will flow along the conductor. This flow is from the negatively charged body to the positively charged body until the two charges are equalized and the potential difference no longer exists.

Note: The basic unit of potential difference is the *volt* (V). The symbol for potential difference is V, indicating the ability to do the work of forcing electrons (current flow) to move. Because the volt unit is used, potential difference is called *voltage*.

WATER ANALOGY

In attempting to train individuals in the concepts of basic electricity, especially in regard to difference of potential (voltage), current, and resistance relationships in a simple electrical circuit, it has been common practice to use what is referred to as the water analogy. We use the water analogy later to explain (in simple straightforward fashion) voltage, current, and resistance and their relationships in more detail, but for now we use the analogy to explain the basic concept of electricity: difference of potential, or voltage. Because a difference in potential causes current flow (against resistance), it is important that this concept be understood first before the concept of current flow and resistance are explained.

Consider the water tanks shown in Figure 13.10—two water tanks connected by a pipe and valve. At first, the valve is closed and all the water is in Tank A. Thus, the water pressure across the valve is at maximum. When the valve is opened, the water flows through the pipe from A to B until the water level becomes the same in both

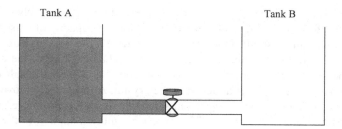

FIGURE 13.10   Water analogy of electric difference of potential

tanks. The water then stops flowing in the pipe, because there is no longer a difference in water pressure (difference in potential) between the two tanks.

Just as the flow of water through the pipe in Figure 13.10 is directly proportional to the difference in water level in the two tanks, current flow through an electric circuit is directly proportional to the difference in potential across the circuit.

Important point: A fundamental law of current electricity is that the current is directly proportional to the applied voltage; that is, if the voltage is increased, the current is increased. If the voltage is decreased, the current is decreased.

Principal Methods of Producing a Voltage

There are many ways to produce electromotive force, or voltage. Some of these methods are much more widely used than others. The following is a list of the seven most common methods of producing electromotive force (USDOE 1992).

a. *Friction*: Voltage produced by rubbing two materials together (static electricity or electrostatic force). Remember our discussion of static electricity? Let's refresh our memories. For example, have you ever walked across a carpet and received a shock when you touched a metal door knob? Your shoe soles built up a charge by rubbing on the carpet, and this charge was transferred to your body. Your body became positively charged and, when you touched the zero-charged door knob, electrons were transferred to your body until both you and the door knob had equal charges.

b. *Pressure (piezoelectricity)*: Voltage produced by squeezing or applying pressure to crystals of certain substances (e.g., crystals like quartz or Rochelle salts or certain ceramics like barium titanate). When pressure is applied to such substances, electrons can be driven out of orbit in the direction of the force. Electrons leave one side of the material and accumulate on the other side, building up positive and negative charges on opposite sides. When the pressure is released, the electrons return to their orbits. Some materials will react to bending pressure, while others will respond to twisting pressure. This generation of voltage is known as the *piezoelectric effect*. If external wires are connected while pressure and voltage are present, electrons will flow and current will be produced. If the pressure is held constant, the current will flow until the potential difference is equalized. When the force is removed, the material is decompressed and immediately causes an electric force in the opposite direction. The power capacity of these materials is extremely small. However, these materials are very useful because of their extreme sensitivity to changes of mechanical force. One example is the crystal phonograph cartridge that contains a Rochelle salt crystal. A phonograph needle is attached to the crystal. As the needle moves in the grooves of a record, it swings from side to side, applying compression and decompression to the crystal. This mechanic motion applied to the crystal generates a voltage signal that is used to reproduce sound.

c. *Heat (thermoelectricity)*: Voltage produced by heating the joint (junction) where two unlike metals are joined. Some materials readily give up their electrons and others readily accept electrons. For example, when two dissimilar metals like copper and zinc are joined together, a transfer of electrons can take place. Electrons will leave the copper atoms and enter the zinc atoms. The zinc gets a surplus of electrons and becomes negatively charged. The copper loses electrons and takes on a positive charge. This creates a voltage potential across the junction of the two metals. The heat energy of normal room temperature is enough to make them release and gain electrons, causing a measurable voltage potential As more heat energy is applied to the junction, more electrons are released, and the voltage potential becomes greater. When heat is removed, the junction cools, the charges will dissipate, and the voltage potential will decrease. This process is called thermoelectricity. A device like this is generally referred to as a thermocouple.

The thermocouple voltage in a thermocouple is dependent upon the heat energy applied to the junction of the two dissimilar metals. Thermocouples are widely used to measure temperature and as heat-sensing devices in automatic temperature-controlled equipment.

Thermocouple power capacities are very small compared to some other sources but are somewhat greater than those of crystals. Generally speaking, a thermocouple can be subjected to higher temperatures than ordinary mercury or alcohol thermometers.

d. *Light (photoelectricity)*: voltage produced by light (photons) striking photosensitive (light sensitive) substances. When the photons in a light beam strike the surface of a material, they release their energy and transfer it to the atomic electrons of the material. This energy transfer may dislodge electrons from their orbits around the surface of the substance. Upon losing electrons, the photosensitive (light sensitive) material becomes positively charged and an electric force is created.

This phenomenon is called the photoelectric effect and has wide applications in electronics, such as photoelectric cells, photovoltaic cells, optical couplers, and television camera tubes. Three uses of the photoelectric effect are described below.

• *Photovoltaic*: The light energy in one of two plates that are joined together causes one plate to release electrons to the other. The plates build up opposite charges, like a battery.

- *Photoemission*: The photon energy from a beam of light could cause a surface to release electrons in a vacuum tube. A plate would then collect the electrons.
- *Photoconduction*: The light energy applied to some materials that are normally poor conductors causes free electrons to be produced in the materials so that they become better conductors.

e. *Chemical action*: Voltage produced by chemical reaction in a battery cell. For example, the voltaic chemical cell wherein a chemical reaction produces and maintains opposite charges on two dissimilar metals that serve as positive and negative terminals. The metals are in contact with an electrolyte solution. Connecting together more than one of these cells will produce a battery.

f. *Magnetism*: Voltage produced in a conductor when the conductor moves through a magnetic field, or a magnetic field moves through the conductor in such a manner as to cut the magnetic lines of force of the field. A generator is a machine that converts mechanical energy into electrical energy by using the principle of *magnetic induction*. This is one of the most useful and widely-employed applications of producing vast quantities of electric power.

g. *Thermionic emission*: A thermionic energy converter is a device consisting of two electrodes placed near one another in a vacuum. One electrode is normally called the cathode, or emitter, and the other is called the anode, or plate. Ordinarily electrons in the cathode are prevented from escaping form the surface by a potential-energy barrier. When an electron starts to move away from the surface, it induces a corresponding positive charge in the material, which tends to pull it back into the surface. To escape, the electron must somehow acquire enough energy to overcome this energy barrier. At ordinary temperatures, almost none of the electrons can acquire enough energy to escape. However, when the cathode is very hot, the electron energies are greatly increased by thermal motion. At sufficiently high temperature, a considerable number of electrons are able to escape. The liberation of electrons from a hot surface is called *thermionic emission*.

The electrons that have escaped from the hot cathode form a cloud of negative charges near it called a space charge. If the plate is maintained positive with respect to the cathode by a battery, the electrons in the cloud are attracted to it. As long as the potential difference between the electrodes is maintained, there will be a steady current flow from the cathode to the plate.

The simplest example of a thermionic device is a vacuum tube diode in which the only electrodes are the cathode and plate, or anode. The diode can be used to convert alternating current (AC) flow to a pulsating direct current (DC) flow.

In the study of the basic electricity related to renewable energy production, we are most concerned with magnetism (generators powered by hydropower, for example), light (photoelectricity produced by solar cells), and chemistry (chemical energy converted to electricity in batteries) as means to produce voltage. Friction has little practical applications, though we discussed it earlier in studying static electricity.

Pressure and heat do have useful applications, but we do not need to consider them in this text. Magnetism used in generators, solar light-produced electricity and the chemistry involved with storing electricity in batteries, on the other hand, are, as mentioned, the principal sources of voltage and are discussed at length in this text.

ELECTRIC CURRENT

The movement or the flow of electrons is called *current*. To produce current, the electrons must be moved by a potential difference or pressure (voltage).

Note: The terms current, current flow, electron flow, or electron current, etc., may be used to describe the same phenomenon.

For our purposes in this text, electron flow, or current, in an electric circuit is from a region of less negative potential to a region of more positive potential—from negative to positive.

Note: Current is represented by the letter I. The basic unit in which current is measured is the *ampere, or amp (A)*. One ampere of current is defined as the movement of one coulomb past any point of a conductor during one second of time.

Recall that we used the water analogy to help us understand potential difference. We can also use the water analogy to help us understand current flow through a simple electric circuit. Consider Figure 13.11 that shows a water tank connected via a pipe to a pump with a discharge pipe. If the water tank contains an amount of water above the level of the pipe opening to the pump, the water exerts pressure (a difference in potential) against the pump. When sufficient water is available for pumping with the pump, water flows through the pipe against the resistance of the pump and pipe. The analogy should be clear—in an electric circuit if a difference of potential exists, current will flow in the circuit.

Another simple way of looking at this analogy is to consider Figure 13.11 where the water tank has been replaced with a generator, the pipe with a conductor (wire) and water flow with the flow of electric current. Figure 13.12 shows a simple electrical circuit.

Again, the key point illustrated by Figures 13.11 and 13.12 is that to produce current, the electrons must be moved by a potential difference.

Electric current is generally classified into two general types:

- direct current (d-c)
- alternating current (a-c)

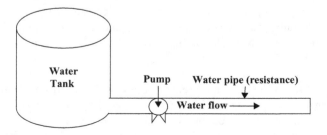

FIGURE 13.11   Water analogy: current flow.

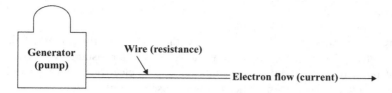

FIGURE 13.12   Simple electric circuit with current flow.

Direct current is current that moves through a conductor or circuit in one direction only. *Alternating current* periodically reverses direction.

RESISTANCE

Earlier, it was pointed out that free electrons, or electric current, could move easily through a good conductor, such as copper, but that an insulator, such as glass, was an obstacle to current flow. In the water analogy shown in Figure 13.11 and the simple electric circuit shown in Figure 13.12, resistance is indicated by either the pipe or the conductor.

Every material offers some resistance, or opposition, to the flow of electric current through it. Good conductors such as copper, silver, and aluminum, offer very little resistance. Poor conductors, or insulators, such as glass, wood, and paper, offer a high resistance to current flow.

Note: The amount of current that flows in a given circuit depends on two factors: voltage and resistance.

Note: Resistance is represented by the letter R. The basic unit in which resistance is measured is the *ohm* (Ω). One ohm is the resistance of a circuit element, or circuit, that permits a steady current of 1 ampere (1 coulomb per second) to flow when a steady electromotive force (emf) of 1 volt is applied to the circuit. Manufactured circuit parts containing definite amounts of resistance are called **resistors**.

The size and type of material of the wires in an electric circuit are chosen so as to keep the electrical resistance as low as possible. In this way, current can flow easily through the conductors, just as water flows through the pipe between the tanks in Figure 13.11. If the water pressure remains constant the flow of water in the pipe will depend on how far the valve is opened. The smaller the opening, the greater the opposition (resistance) to the flow, and the smaller will be the rate of flow in gallons per second.

In the electric circuit shown in Figure 13.12, the larger the diameter of the wire, the lower will be its electrical resistance (opposition) to the flow of current through it. In the water analogy, pipe friction opposes the flow of water between the tanks. This friction is similar to electrical resistance. The resistance of the pipe to the flow of water through it depends upon (1) the length of the pipe, (2) diameter of the pipe, and (3) the nature of the inside walls (rough or smooth). Similarly, the electrical resistance of the conductors depends upon (1) the length of the wires, (2) the diameter of the wires, and (3) the material of the wires (copper, silver, etc.).

It is important to note that temperature also affects the resistance of electrical conductors to some extent. In most conductors (copper, aluminum, etc.) the resistance increases with temperature. Carbon is an exception. In carbon the resistance decreases as temperature increases.

Important note: Electricity is a study that is frequently explained in terms of opposites. The term that is exactly the opposite of resistance is *conductance*. Conductance (G) is the ability of a material to pass electrons. The SI derived unit of conductance is the siemens. The commonly used unit of conductance is the *Mho*, which is ohm spelled backward. The relationship that exists between resistance and conductance is the reciprocal. A reciprocal of a number is obtained by dividing the number into one. If the resistance of a material is known, dividing its value into one will give its conductance. Similarly, if the conductance is known, dividing its value into one will give its resistance.

ELECTROMAGNETISM

Earlier, fundamental theories concerning simple magnets and magnetism were presented. Those discussions dealt mainly with forms of magnetism that were not related directly to electricity—permanent magnets for instance. Further, only brief mention was made of those forms of magnetism having direct relation to electricity—producing electricity with magnetism for instance.

In medicine, anatomy and physiology are so closely related that the medical student cannot study one at length without involving the other. A similar relationship holds for the electrical field; that is, magnetism and basic electricity are so closely related that one cannot be studied at length without involving the other. This close fundamental relationship is continually borne out in the study of generators, transformers, battery packs, and motors. To be proficient in electricity, we must become familiar with such general relationships that exist between magnetism and electricity as follows:

a. Electric current flow will always produce some form of magnetism.
b. Magnetism is by far the most commonly used means for producing or using electricity.
c. The peculiar behavior of electricity under certain conditions is caused by magnetic influences.

MAGNETIC FIELD AROUND A SINGLE CONDUCTOR

In 1819, Hans Christian Oersted, a Danish scientist, discovered that a field of magnetic force exists around a single wire conductor carrying an electric current. In Figure 13.13, a wire is passed through a piece of cardboard and connected through a switch to a dry cell. With the switch open (no current flowing), if we sprinkle iron filings on the cardboard then tap it gently, the filings will fall back haphazardly. Now, if we close the switch, current will begin to flow in the wire. If we tap the cardboard again, the magnetic effect of the current in the wire will cause the filings to fall back into a definite pattern of concentric circles with the wire as the center of the circles. Every section of the wire has this field of force around it in a plane perpendicular to the wire, as shown in Figure 3.14.

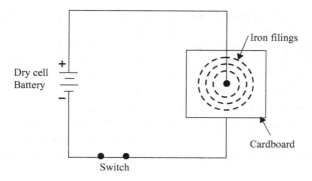

FIGURE 13.13   A circular pattern of magnetic force exists around a wire carrying an electric current.

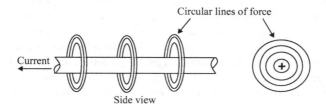

FIGURE 13.14   The circular fields of force around a wire carrying a current are in planes which are perpendicular to the wire.

FIGURE 13.15   The strength of the magnetic field around a wire carrying a current depends on the amount of current.

The ability of the magnetic field to attract bits of iron depends on the number of lines of force present. The strength of the magnetic field around a wire carrying a current depends on the current, since it is the current that produces the field. The greater the current, the greater the strength of the field. A large current will produce many lines of force extending far from the wire, while a small current will produce only a few lines close to the wire, as shown in Figure 13.15.

POLARITY OF A SINGLE CONDUCTOR

The relation between the direction of the magnetic lines of force around a conductor and the direction of current flow along the conductor may be determined by means of the **left-hand rule for a conductor**. If the conductor is grasped in the left hand with the thumb extended in the direction of electron flow (- to +), the fingers will point in the direction of the magnetic lines of force. This is the same direction that the north pole of a compass would point if the compass were placed in the magnetic field.

Important note: Arrows are generally used in electric diagrams to denote the direction of current flow along the length of wire. Where cross sections of wire are shown, a special view of the arrow is used. A cross-sectional view of a conductor that is carrying current toward the observer is illustrated in Figure 13.16a. The direction of current is indicated by a dot, which represents the head of the arrow. A conductor that is carrying current away from the observer is illustrated in Figure 13.16b. The direction of current is indicated by a cross, which represents the tail of the arrow.

FIELD AROUND TWO PARALLEL CONDUCTORS

When two parallel conductors carry current in the same direction, the magnetic fields tend to encircle both conductors, drawing them together with a force of attraction, as shown in Figure 13.17a. Two parallel conductors carrying currents in opposite directions are shown in Figure 13.17b. The field around one conductor is opposite in direction to the field around the other conductor. The resulting lines of force are crowded together in the space between the wires and tend to push the wires apart. Therefore, two parallel adjacent conductors carrying currents in the same direction attract each other and two parallel conductors carrying currents in opposite directions repel each other.

MAGNETIC FIELD OF A COIL

The magnetic field around a current-carrying wire exists at all points along its length. Bending the current-carrying wire into the form of a single loop has two results. First, the magnetic field consists of more dense concentric circles in a plane perpendicular to the wire (see Figure 13.18), although the total number of lines is the same as for the straight conductor. Second, all the lines inside the loop are in the same direction. When this straight wire is wound around a core, as is shown in Figure 13.20), it becomes a coil and the magnetic field assumes a different shape. When current is passed through the coiled conductor, the magnetic field of each turn of wire links

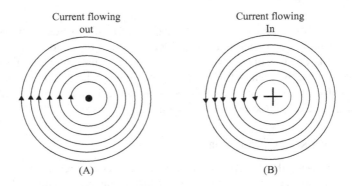

FIGURE 13.16   Magnetic field around a current-carrying conductor.

(A) Current flowing in the same direction

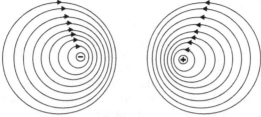

(B) Currents flowing in opposite directions

FIGURE 13.17 Magnetic field around two parallel conductors.

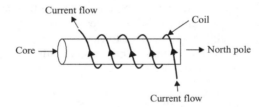

FIGURE 13.18 Current-carrying coil.

with the fields of adjacent turns. The combined influence of all the turns produces a two-pole field similar to that of a simple bar magnet. One end of the coil will be a north pole and the other end will be a south pole.

POLARITY OF AN ELECTROMAGNETIC COIL

In Figure 13.17, it was shown that the direction of the magnetic field around straight conductor depends on the direction of current flow through that conductor. Thus, a reversal of current flow through a conductor causes a reversal in the direction of the magnetic field that is produced. It follows that a reversal of the current flow through a coil also causes a reversal of its two-pole field. This is true because that field is the product of the linkage between the individual turns of wire on the coil. Therefore, if the field of each turn is reversed, it follows that the total field (coils' field) is also reversed.

When the direction of electron flow through a coil is known, its polarity may be determined by use of the **left-hand rule for coils**. This rule is illustrated in Figure 13.18 and is stated as follows: Grasping the coil in the left hand, with the fingers "wrapped around" in the direction of electron flow, the thumb will point toward the North Pole.

STRENGTH OF AN ELECTROMAGNETIC FIELD

The strength, or intensity, of the magnetic field of a coil depends on a number of factors.

- The *number of turns* of conductor.
- The *amount of current flow* through the coil.
- The *ratio of the coil's length to its width*.
- The *type of material in the core*.

MAGNETIC UNITS

The law of current flow in the electric circuit is similar to the law for the establishing of flux in the magnetic circuit.

The *magnetic flux*, φ, (phi) is similar to current in the Ohm's Law formula and comprises the total number of lines of force existing in the magnetic circuit. The **Maxwell** is the unit of flux—that is, 1 line of force is equal to 1 maxwell.

Note: The maxwell is often referred to as simply a line of force, line of induction, or line.

The *strength* of a magnetic field in a coil of wire depends on how much current flows in the turns of the coil. The more current, the stronger the magnetic field. Also, the more turns, the more concentrated are the lines of force. The *force* that produces the flux in the magnetic circuit (comparable to electromotive force in Ohm's Law) is known as *magnetomotive force*, or mmf. The practical unit of magnetomotive force is the **ampere-turn** (At). In equation form,

$$F = \text{ampere} - \text{turns} = NI \qquad (13.4)$$

where

F = magnetomotive force, At
N = number of turns
I = current, A

Example 13.5

Problem:

Calculate the ampere-turns for a coil with 2,000 turns and a 5-Ma current.

Solution:

Use equation (13.6) and substitute N = 2,000 and I = 5 x 10⁻³ A.

$$NI = 2,000\left(5 \times 10^{-3}\right) = 10 \text{ At}$$

The unit of *intensity* of magnetizing force per unit of length is designated as H and is sometimes expressed as Gilberts per centimeter of length. Expressed as an equation,

$$H = \frac{NI}{L} \tag{13.5}$$

where
  H = magnetic field intensity, ampere-turns per meter (At/m)
  NI = ampere-turns, At
  L = length between poles of the coil, m

Note: Equation (13.5) is for a solenoid. H is the intensity of an air core. With an iron core, H is the intensity through the entire core and L is the length or distance between poles of the iron core.

PROPERTIES OF MAGNETIC MATERIALS

In this section, we discuss two important properties of magnetic materials: permeability and hysteresis.

Permeability

When the core of an electromagnet is made of annealed sheet steel it produces a stronger magnet than if a cast iron core is used. This is the case because annealed sheet steel is more readily acted upon by the magnetizing force of the coil than is the hard cast iron. Simply put, soft sheet steel is said to have greater *permeability* because of the greater ease with which magnetic lines are established in it.

Recall that permeability is the relative ease with which a substance conducts magnetic lines of force. The permeability of air is arbitrarily set at 1. The permeability of other substances is the ratio of their ability to conduct magnetic lines compared to that of air. The permeability of nonmagnetic materials, such as aluminum, copper, wood, and brass is essentially unity, or the same as for air.

Important note: The permeability of magnetic materials varies with the degree of magnetization, being smaller for high values of flux density. *Reluctance*—which is analogous to resistance is the opposition to the production of flux in a material—is inversely proportional to permeability. Iron has high permeability and, therefore, low reluctance. Air has low permeability and hence high reluctance.

Hysteresis

When the current in a coil of wire reverses thousands of times per second, a considerable loss of energy can occur. This loss of energy is caused by *hysteresis*. Hysteresis means "a lagging behind"; that is, the magnetic flux in an iron core lags behind the increases or decreases of the magnetizing force. The simplest method of illustrating the property of hysteresis is by graphical means such as the hysteresis loop shown in Figure 13.19.

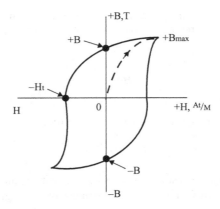

FIGURE 13.19   Hysteresis loop.

HYSTERESIS LOOP

The hysteresis loop (Figure 13.19) is a series of curves that show the characteristics of a magnetic material. Opposite directions of current result are in the opposite directions of $+H$ and $-H$ for field intensity. Similarly, opposite polarities are shown for flux density as $+B$ and $-B$. The current starts at the center 0 (zero) when the material is unmagnetized. Positive H values increase B to saturation at $+B_{max}$. Next H decreases to zero, but B drops to the value of B, because of hysteresis. The current that produced the original magnetization now is reversed so that H becomes negative. B drops to zero and continues to $-B_{max}$. As the $-H$ values decrease, B is reduced to $-B$, when H is zero. Now with a positive swing of current, H becomes positive, producing saturation at $+B_{max}$ again. The hysteresis loop is now completed. The curve doesn't return to zero at the center because of hysteresis.

ELECTROMAGNETS

An *electromagnet* is composed of a coil of wire wound around a core that is normally soft iron, because of its high permeability and low hysteresis. When direct current flows through the coil, the core will become magnetized with the same polarity that the coil would have without the core. If the current is reversed, the polarity of the coil and core are reversed.

The electromagnet is of great importance in electricity simply because the magnetism can be "turned on" or "turned off" at will. The starter solenoid (an electromagnet) in automobiles and power boats is a good example. In an automobile or boat, an electromagnet is part of a relay that connects the battery to the induction coil, which generates the very high voltage needed to start the engine. The starter solenoid isolates this high voltage from the ignition switch. When no current flows in the coil, it is an "air core," but when the coil is energized, a movable soft-iron core does two things. First, the magnetic flux is increased because the soft-iron core is more permeable than the air core. Second, the flux is more highly concentrated.

All this concentration of magnetic lines of force in the soft-iron core results in a very good magnet when current flows in the coil. But soft-iron loses its magnetism quickly when the current is shut off. The effect of the soft iron is, of course, the same whether it is movable, as in some solenoids, or permanently installed in the coil. An electromagnet, then consists basically of a coil and a core; it becomes a magnet when current flows through the coil.

The ability to control the action of magnetic force makes an electromagnet very useful in many circuit applications. Many of the applications of electromagnets are discussed throughout this manual.

Note: In order to understand electricity (current flow in a conductor) and having a feel, so to speak, of what is going on whenever electricity is flowing in a circuit, simple electrical circuit math examples are provided in the next chapter.

NOTE

1 Much of the information in this section is adapted from F.R. Spellman (2001, 2022).

REFERENCES

Spellman, F.R. (2001). *Electricity*. Boca Raton, FL, CRC Press.
Spellman, F.R. (2022). *The science of wind power*. Boca Raton, FL, CRC Press.

14 Electrical Calculations

INTRODUCTION

This chapter provides an in-depth discussion of typical direct current (DC) electrical calculations because it is important to understand electricity when dealing with electrical vehicle operations; therefore, practical (very basic) electrical problems are presented. Alternating current (AC) is not presented here because electrical knowledge beyond Ohm's Law and basic circuit analysis is all that is required. Keep in mind that all only the very basics are presented herein. It should be pointed out, however, that only "qualified" electricians should perform electrical work of any type.

ELECTRICAL CALCULATIONS

An electric circuit includes: an **energy source** [source of electromotive force (emf) or voltage; that is, a battery or generator], a conductor (wire), a load, and a means of control (see Figure 14.1). The energy source could be a battery, as in Figure 14.1, or some other means of producing a voltage. The **load** that dissipates the energy could be a lamp, a resistor, or some other device (or devices) that does useful work, such as an electric toaster, a power drill, radio, or a soldering iron. **Conductors** are wires that offer low resistance to current; they connect all the loads in the circuit to the voltage source. No electrical device dissipates energy unless current flows through it. Since conductors, or wires, are not perfect conductors, they heat up (dissipate energy), so they are actually part of the load. For simplicity, however, we usually think of the connecting wiring as having no resistance, since it would be tedious to assign a very low resistance value to the wires every time we wanted

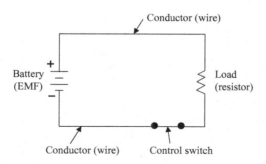

FIGURE 14.1 Simple closed circuit.

DOI: 10.1201/9781003387879-18

to solve a problem. **Control devices** might be switches, variable resistors, circuit breakers, fuses, or relays.

OHM'S LAW

Simply put, *Ohm's Law* defines the relationship between current, voltage, and resistance in electric circuits. Ohm's Law can be expressed mathematically in three ways.

a. The *current* in a circuit is equal to the voltage applied to the circuit divided by the resistance of the circuit. Stated another way, the current in a circuit is DIRECTLY proportional to the applied voltage and INVERSELY proportional to the circuit resistance. Ohm's Law may be expressed as an equation:

$$I = \frac{E}{R} \qquad (14.1)$$

where
I = current in amperes (amps)
E = voltage in volts
R = resistance in ohms

b. The *resistance* of a circuit is equal to the voltage applied to the circuit divided by the current in the circuit:

$$R = \frac{E}{I} \qquad (14.2)$$

c. The applied *voltage* (E) to a circuit is equal to the product of the current and the resistance of the circuit:

$$E = I \times R = IR \qquad (14.3)$$

If any two of the quantities in equations (14.1)–(14.3) are known, the third may be easily found. Let's look at an example.

Example 14.1

Problem:

Figure 10.2 shows a circuit containing a resistance of 6Ω and a source of voltage of 3 V. How much current flows in the circuit?
Given:

$$E = 3 \text{ V}$$

$$R = 6\Omega$$

$$I = ?$$

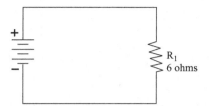

FIGURE 14.2   Determining current in a simple circuit.

Solution:

$$I = \frac{E}{R}$$

$$I = \frac{3}{6}$$

$$I = 0.5 \text{ amperes}$$

To observe the effect of source voltage on circuit current, we use the circuit shown in Figure 14.2 but double the voltage to 6 V.

Example 14.2

Problem:

$$E = 3 \text{ V}$$
$$R = 6\Omega$$
$$I = ?$$

Solution:

$$I = \frac{E}{R}$$

$$I = \frac{6}{6}$$

$$I = 1 \text{ ampere}$$

Notice that as the source of voltage doubles, the circuit current also doubles.

The key point: Circuit current is directly proportional to applied voltage and will change by the same factor that the voltage changes.

To verify that current is inversely proportional to resistance, assume the resistor in Figure 14.2 to have a value of 12Ω.

Example 14.3

Problem:

Given:

$$E = 3\,\text{V}$$
$$R = 12\,\Omega$$
$$I = ?$$

Solution:

$$I = \frac{E}{R}$$
$$I = \frac{3}{12}$$
$$I = 0.25 \text{ amperes}$$

Comparing this current of 0.25 ampere for the 12-Ω resistor, to the 0.5-ampere of current obtained with the 6-Ω resistor, shows that doubling the resistance will reduce the current to one half the original value. The point is that circuit current is inversely proportional to the circuit resistance.

Recall that if you know any two quantities E, I, and R, you can calculate the third. In many circuit applications, current is known and either the voltage or the resistance will be the unknown quantity. To solve a problem, in which current and resistance are known, the basic formula for Ohm's Law must be transposed to solve for E, for I, or for R. However, the Ohm's Law equations can be memorized and practiced effectively by using an Ohm's Law circle (see Figure 14.3). To find the equation for E, I, or R when two quantities are known, cover the unknown third quantity with your finger, ruler, piece of paper, etc., as shown in Figure 14.4.

Example 14.4

Problem:

Find I when $E = 120$ V and $R = 40\Omega$.

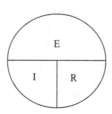

FIGURE 14.3   Ohm's Law circle.

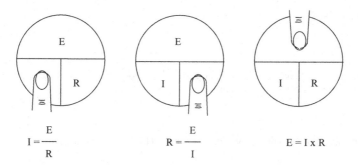

$$I = \frac{E}{R}$$ $$R = \frac{E}{I}$$ $$E = I \times R$$

FIGURE 14.4   Putting Ohm's Law circle to work.

Solution:

Place finger on *I* as shown in Figure 14.4. Use equation (14.1) to find the unknown *I*.

Example 14.5

Problem:

Find *R* when *E* = 220 V and *I* = 10 ampere

Solution:

Place finger on *R* as shown in Figure 14.4.
   Use equation (14.2) to find the unknown *R*.
   220 V/10 ampere = 22Ω

Example 14.6

Problem:

Find *E* when *I* = 2.5 ampere and *R* = 25Ω.

Solution:

$$E = IR = 2.5(25) = 62.5 \text{ V}$$

Note: In the previous examples, we have demonstrated how the Ohm's Law circle can help us solve simple voltage, current and amperage problems. The beginning student is cautioned, however, not to rely wholly on the use of this circle when transposing simple formulas but rather to use it to supplement his/her knowledge of the algebraic method. Algebra is a basic tool in the solution of electrical problems and the importance of knowing how to use it should not be under-emphasized or bypassed after the operator has learned a shortcut method such as the one indicated in this circle.

Example 14.7

Problem:

An electric light bulb draws 0.5 ampere when operating on a 120-V DC circuit. What is the resistance of the bulb?

Solution:

Because E and I are known, we use equation (14.2) to solve for R.

$$R = \frac{E}{I} = \frac{120}{0.5} = 240\Omega$$

ELECTRIC POWER

Power, whether electrical or mechanical, pertains to the rate at which work is being done, so the power consumption in your plant is related to current flow. A large electric motor or air dryer consumes more power (and draws more current) in a given length of time than, for example, an indicating light on a motor controller. **Work** is done whenever a force causes motion. If a mechanical force is used to lift or move a weight, work is done. However, force exerted WITHOUT causing motion, such as the force of a compressed spring acting between two fixed objects, does not constitute work.

Key points: Power is the rate at which work is done.

ELECTRICAL POWER CALCULATIONS

The electric power P used in any part of a circuit is equal to the current I in that part multiplied by the V across that part of the circuit. In equation form,

$$P = EI \tag{14.4}$$

where
  P = power, Watts (W)
  E = voltage, V
  I = current, A

If we know the current I and the resistance R but not the voltage V, we can find the power P by using Ohm's Law for voltage, so that substituting

$$E = IR$$

$$P = IR \times I = I^2R \tag{14.5}$$

In the same manner, if we know the voltage V and the resistance R but not the current I, we can find the P by using Ohm's Law for current, so that substituting

$$I = \frac{E}{R}$$

$$P = E = \frac{E}{R} = \frac{E^2}{R} \qquad (14.6)$$

Key point: If we know any two quantities, we can calculate the third.

Example 14.8

Problem:

The current through a 200-Ω resistor to be used in a circuit is 0.25 ampere. Find the power rating of the resistor.

Solution:

Because R and I are known, use equation (14.5) to find p.

$$P = I^2R = (0.25)^2 (200) = 0.0625(200) = 12.5 \text{ W}$$

Example 14.9

Problem:

How many kilowatts of power are delivered to a circuit by a 220-V generator that supplies 30 ampere to the circuit?

Solution:

Because E and I are given, use equation (14.4) to find P.

$$P = EI = 220(30) = 6,600 \text{ W} = 6.6 \text{ Kw}$$

Example 14.10

Problem:

If the voltage across a 30,000-Ω resistor is 450 V, what is the power dissipated in the resistor?

Solution:

Because R and E are known, use equation (14.6) to find P.

$$P = \frac{E^2}{R} = \frac{450^2}{30,000} = \frac{202,500}{30,000} = 6.75 \text{ W}$$

In this section, *P* was expressed in terms of alternate pairs of the other three basic quantities *E*, *R*, and *I*. In practice, you should be able to express any one of the three basic quantities, as well as *P*, in terms of any two of the others.

ELECTRIC ENERGY

Energy (the mechanical definition) is defined as the ability to do work (energy and time are essentially the same and are expressed in identical units). Energy is expended when work is done, because it takes energy to maintain a force when that force acts through a distance. The total energy expended to do a certain amount of work is equal to the working force multiplied by the distance through which the force moved to do the work.

In electricity, total energy expended is equal to the *rate* at which work is done, multiplied by the length of time the rate is measured. Essentially, energy W is equal to power P times time t. The kilowatt-hour (kWh) is a unit commonly used for large amounts of electric energy or work. The amount of kilowatt-hours is calculated as the product of the power in kilowatts (Kw) and the time in hours (h) during which the power is used.

$$kWh = Kw \times h \qquad\qquad (14.7)$$

Example 14.11

Problem:

How much energy is delivered in 4 hours by a generator supplying 12 Kw?

Solution:

$$kWh = Kw \times h = 12(4) = 48$$

Energy delivered = 48 kWh.

SERIES DC CIRCUIT CHARACTERISTICS

As previously mentioned, an electric circuit is made up of a voltage source, the necessary connecting conductors, and the effective load. If the circuit is arranged so that the electrons have only ONE possible path, the circuit is called a *Series Circuit*. Therefore, a series circuit is defined as a circuit that contains only one path for current flow. Figure 14.5 shows a series circuit having several loads (resistors).

Key point: A *series circuit* is a circuit in which there is only one path for current to flow along.

RESISTANCE

Referring to Figure 14.5, the current in a series circuit, in completing its electrical path, must flow through each resistor inserted into the circuit. Thus, each additional

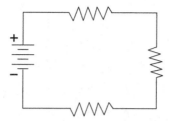

FIGURE 14.5   Series circuit.

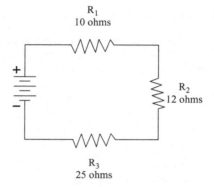

FIGURE 14.6   Solving for total resistance in a series circuit.

resistor offers added resistance. In a series circuit, *the total circuit resistance (R_T) is equal to the sum of the individual resistances.* As an equation:

$$R_T = R_1 + R_2 + R_3 \ldots R_n \tag{14.8}$$

where
  R_T = total resistance, Ω
  R_1, R_2, R_3 = resistance in series, Ω
  R_n = any number of additional resistors in equation

Example 14.12

Problem:

Three resistors of 10Ω, 12Ω, and 25Ω are connected in series across a battery whose emf is 110 V (Figure 14.6). What is the total resistance?

Solution:

Given:

$$R_1 = 10\Omega$$
$$R_2 = 12\Omega$$
$$R_3 = 25\Omega$$
$$R_T = ?$$
$$R_T = R_1 + R_2 + R_3$$
$$R_T = 10 + 12 + 25$$
$$R_T = 47\Omega$$

Equation 14.8 can be transposed to solve for the value of an unknown resistance. For example, transposition can be used in some circuit applications where the total resistance is known but the value of a circuit resistor has to be determined.

Example 14.13

Problem:

The total resistance of a circuit containing three resistors is 50Ω (see Figure 14.7). Two of the circuit resistors are 12Ω each. Calculate the value of the third resistor.

Solution:

Given:

$$R_T = 50\Omega$$
$$R_1 = 50\Omega$$
$$R_2 = 12\Omega$$

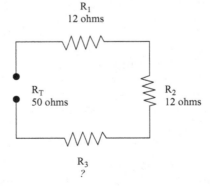

FIGURE 14.7   For Example 14.13.

$$R_3 = ?$$
$$R_T = R_1 + R_2 + R_3$$

Subtracting $(R_1 + R_2)$ from both sides of the equation

$$R_3 = R_T - R_1 - R_2$$
$$R_3 = 50 - 12 - 12$$
$$R_3 = 50 - 24$$
$$R_3 = 26\Omega$$

Key point: When resistances are connected in series, the total resistance in the circuit is equal to the sum of the resistances of all the parts of the circuit.

CURRENT

Because there is but one path for current in a series circuit, the same *current* (I) must flow through each part of the circuit. Thus, to determine the current throughout a series circuit, only the current through one of the parts need be known. The fact that the same current flows through each part of a series circuit can be verified by inserting ammeters into the circuit at various points as shown in Figure 14.8. As indicated in Figure 14.8, each meter indicates the same value of current.

   Key point: In a series circuit, the same current flows in every part of the circuit. *Do not* add the currents in each part of the circuit to obtain *I*.

VOLTAGE

The *voltage* drop across the resistor in the basic circuit is the total voltage across the circuit and is equal to the applied voltage. The total voltage across a series circuit is

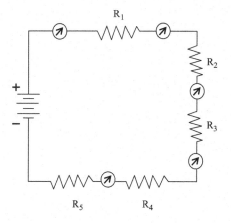

FIGURE 14.8 Current in a series circuit.

also equal to the applied voltage but consists of the sum of two or more individual voltage drops. This statement can be proven by an examination of the circuit shown in Figure 14.9. In this circuit, a source potential (E_T) of 30 V is impressed across a series circuit consisting of two 6-Ω resistors. The total resistance of the circuit is equal to the sum of the two individual resistances, or 12Ω. Using Ohm's Law, the circuit current may be calculated as follows:

$$I = \frac{E_T}{R_T}$$

$$I = \frac{30}{12}$$

$$I = 2.5 \text{ amperes}$$

Knowing the value of the resistors to be 6Ω each, and the current through the resistors to be 2.5 amperes, the voltage drops across the resistors can be calculated. The voltage (E_1) across R_1 is therefore:

$$E_1 = IR_1$$

$$E_1 = 2.5 \text{ amperes} \times 6\Omega$$

$$E_1 = 15 \text{ V}$$

Since R_2 is the same ohmic value as R_1 and carries the same current, the voltage drop across R_2 is also equal to 15 V. Adding these two 15 V drops together gives a total drop of 30 V exactly equal to the applied voltage. For a series circuit then:

$$E_T = E_1 + E_2 + E_3 \dots E_n \tag{14.9}$$

where
E_T = total voltage, V
E_1 = voltage across resistance R_1, V

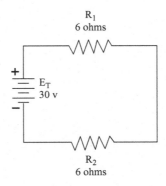

FIGURE 14.9 Calculating total resistance in a series circuit.

E_2 = voltage across resistance R_2, V
E_3 = voltage across resistance R_3, V

Example 14.14

Problem:

A series circuit consists of three resistors having values of 10Ω, 20Ω, and 40Ω respectively. Find the applied voltage if the current through the 20-Ω resistor is 2.5 amperes.

Solution:

To solve this problem, a circuit diagram is first drawn and labeled as shown in Figure 14.10.
   Given:

$$R_1 = 10\Omega$$

$$R_2 = 20\Omega$$

$$R_3 = 40\Omega$$

$$I = 2.6 \text{ amperes}$$

Because the circuit involved is a series circuit, the same 2.5 amperes of current flows through each resistor. Using Ohm's Law, the voltage drops across each of the three resistors can be calculated and are:

$$E_1 = 25 \text{ V}$$

$$E_2 = 50 \text{ V}$$

$$E_3 = 100 \text{ V}$$

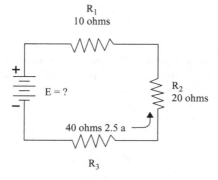

FIGURE 14.10   Solving for applied voltage in a series circuit.

Once the individual drops are known, they can be added to find the total or applied voltage using equation (14.9):

$$E_T = E_1 + E_2 + E_3$$
$$E_T = 25 \text{ V} + 50 \text{ V} + 100 \text{ V}$$
$$E_T = 175 \text{ V}$$

Key point 1: The total voltage (E_T) across a series circuit is equal to the sum of the voltages across each resistance of the circuit.

    Key point 2: The voltage drops that occur in a series circuit are in direct proportions to the resistance across which they appear. This is the result of having the same current flow through each resistor. Thus, the larger the resistor the larger will be the voltage drop across it.

POWER

Each resistor in a series circuit consumes *power*. This power is dissipated in the form of heat. Because this power must come from the source, the total power must be equal in amount to the power consumed by the circuit resistances. In a series circuit, the total power is equal to the **sum** of the powers dissipated by the individual resistors. Total power (P_T) is thus equal to:

$$P_T = P_1 + P_2 + P_3 ... P_n \tag{14.10}$$

where
    P_T = total power, W
    P_1 = power used in first part, W
    P_2 = power used in second part, W
    P_3 = power used in third part, W
    P_n = power used in nth part, W

Example 14.15

Problem:

A series circuit consists of three resistors having values of 5Ω, 15Ω, and 20Ω. Find the total power dissipation when 120 V is applied to the circuit (See Figure 14.11).

Solution:

Given:

$$R_1 = 5\Omega$$
$$R_2 = 15\Omega$$

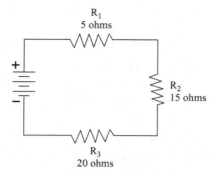

FIGURE 14.11 Solving for total power in a series circuit.

$$R_3 = 20\Omega$$

$$E = 120 \text{ V}$$

The total resistance is found first.

$$R_T = R_1 + R_2 + R_3$$

$$R_T = 5 + 15 + 20$$

$$R_T = 40\Omega$$

Using total resistance and the applied voltage, the circuit current is calculated.

$$I = \frac{E_T}{R_T}$$

$$I = \frac{120}{40}$$

$$I = 3 \text{ amperes}$$

Using the power formula, the individual power dissipations can be calculated.
For resistor R_1:

$$P_1 = I^2 R_1$$

$$P_1 = (3)^2 5$$

$$P_1 = 45 \text{ W}$$

For R_2:

$$P_2 = I^2 R_2$$

$$P_2 = (3)^2 5$$

$$P_2 = 135 \text{ W}$$

For R_3:

$$P_3 = I^2 R_3$$
$$P_3 = (3)^2 20$$
$$P_3 = 180 \text{ W}$$

To obtain total power:

$$P_T = P_1 + P_2 + P_3$$
$$P_T = 45 + 135 + 180$$
$$P_T = 360 \text{ W}$$

To check the answer the total power delivered by the source can be calculated:

$$P = E \times I$$
$$P = 3 \text{ A} \times 120 \text{ V}$$
$$P = 360 \text{ W}$$

Thus the total power is equal to the sum of the individual power dissipations.

Key point: We found that Ohm's Law can be used for total values in a series circuit as well as for individual parts of the circuit. Similarly, the formula for power may be used for total values.

$$P_T = IE_T$$

GENERAL SERIES CIRCUIT ANALYSIS

Now that we have discussed the pieces involved in putting together the puzzle for solving series circuit analysis, we now move on to the next step in the process: solving series circuit analysis in total.

Example 14.16

Problem:

Three resistors of 20Ω, 20Ω, and 30Ω are connected across a battery supply rated at 100 V terminal voltage. Completely solve the circuit shown in Figure 14.12.

Note: In solving the circuit, the total resistance will be found first. Next, the circuit current will be calculated. Once the current is known, the voltage drops and power dissipations can be calculated.

Solution:

The total resistance is:

$$R_T = R_1 + R_2 + R_3$$

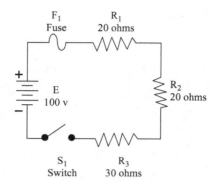

FIGURE 14.12 Solving for various values in a series circuit.

$$R_T = 20\Omega + 20\Omega + 30\Omega$$
$$R_T = 70\Omega$$

By Ohm's Law the current is:

$$I = \frac{E}{R_T}$$
$$I = \frac{100}{70}$$
$$I = 1.43 \text{ amperes (rounded)}$$

The voltage (E_1) across R_1 is:

$$E_1 = IR_1$$
$$E_1 = 1.43 \text{ amperes} + 20\Omega$$
$$E_1 = 28.6 \text{ V}$$

The voltage (E_2) across R_2 is:

$$E_2 = IR_2$$
$$E_2 = 1.43 \text{ amperes} + 20\Omega$$
$$E_2 = 28.6 \text{ V}$$

The voltage (E_3) across R_3 is:

$$E_3 = IR_2$$
$$E_3 = 1.43 \text{ amperes} + 30\Omega$$
$$E_3 = 42.9 \text{ V}$$

The power dissipated by R_1 is:

$$P_1 = I \times E_1$$
$$P_1 = 1.43 \text{ amperes} + 28.6 \text{ V}$$
$$P_1 = 40.9 \text{ W}$$

The power dissipated by R_2 is:

$$P_2 = I \times E_2$$
$$P_2 = 1.43 \text{ amperes} + 28.6 \text{ V}$$
$$P_2 = 40.9 \text{ W}$$

The power dissipated by R_3 is:

$$P_3 = I \times E_3$$
$$P_3 = 1.43 \text{ amperes} + 42.9 \text{ V}$$
$$P_3 = 61.3 \text{ W (rounded)}$$

The total power dissipated is:

$$P_T = E_T \times I$$
$$P_T = 100 \text{ V} \times 1.43 \text{ amperes}$$
$$P_T = 143 \text{ W}$$

Important note: Keep in mind when applying Ohm's Law to a series circuit to consider whether the values used are component values or total values. When the information available enables the use of Ohm's Law to find total resistance, total voltage, and total current, total values must be inserted into the formula.

To find total resistance:

$$R_T = \frac{E_T}{I_T}$$

To find total voltage:

$$E_T = I_T \times R_T$$

To find total current:

$$I_T = \frac{E_T}{R_T}$$

PARALLEL DC CIRCUITS

The principles we applied to solving simple series circuit calculations for determining the reactions of such quantities as voltage, current, and resistance also can be used in parallel and series-parallel circuits.

PARALLEL CIRCUIT CHARACTERISTICS

A *parallel circuit* is defined as one having two or more components connected across the same voltage source (see Figure 14.13). Recall that in a series circuit there is only one path for current flow. As additional loads (resistors, etc.) are added to the circuit, the total resistance increases and the total current decreases. This is *not the case* in a **parallel** circuit. In a parallel circuit, each load (or branch) is connected directly across the voltage source. In Figure 14.13, commencing at the voltage source (E_b) and tracing counterclockwise around the circuit, two complete and separate paths can be identified in which current can flow. One path is traced from the source through resistance R_1 and back to the source; the other, from the source through resistance R_2 and back to the source.

VOLTAGE IN PARALLEL CIRCUITS

Recall that in a series circuit the source voltage divides proportionately across each resistor in the circuit. In a parallel circuit (see Figure 14.13), the same voltage is present across all the resistors of a parallel group. This voltage is equal to the applied voltage (E_b) and can be expressed in equation form as:

$$E_h = E_{R1} = E_{R2} = E_{Rn} \qquad (14.11)$$

We can verify equation (14.11) by taking voltage measurements across the resistors of a parallel circuit, as illustrated in Figure 14.4. Notice that each voltmeter indicates

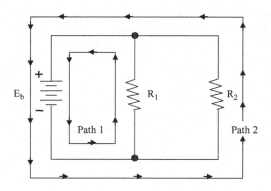

FIGURE 14.13   Basic parallel circuit.

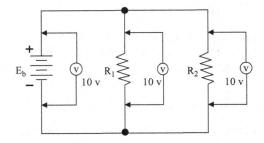

FIGURE 14.14   Voltage comparison in a parallel circuit.

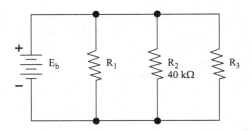

FIGURE 14.15   Refers to Example 14.17.

the same amount of voltage; that is, the voltage across each resistor is the same as the applied voltage.

Key point: In a parallel circuit, the voltage remains the same throughout the circuit.

Example 14.17

Problem:

Assume that the current through a resistor of a parallel circuit is known to be 4.0 milliamperes (ma) and the value of the resistor is 40,000Ω. Determine the potential (voltage) across the resistor. The circuit is shown in Figure 14.15.

Solution:

Given:

$$R_2 = 40 \text{ k}\Omega$$

$$I_{R2} = 4.0 \text{ ma}$$

Find:

$$E_{R2} = ?$$
$$E_h = ?$$

Select proper equation

$$E = IR$$

Substitute known values:

$$E_{R2} = I_{R2} \times R_2$$
$$E_{R2} = 4.0 \text{ ma} \times 40{,}000\Omega$$
$$[\text{use power of tens}]$$
$$E_{R2} = \left(4.0 \times 10^{-3}\right) \times \left(40 \times 10^{3}\right)$$
$$E_{R2} = 4.0 \times 4.0$$

Resultant:

$$E_{R2} = 160 \text{ V}$$

Therefore:

$$E_h = 160 \text{ V}$$

CURRENT IN PARALLEL CIRCUITS

Important point: Ohm's Law states: The current in a circuit is inversely proportional to the circuit resistance. This fact, important as a basic building block of electrical theory, obviously, is also important in the following explanation of current flow in parallel circuits.

In a series circuit, a single current flows. Its value is determined in part by the total resistance of the circuit. However, the source current in a parallel circuit divides among the available paths in relation to the value of the resistors in the circuit. Ohm's Law remains unchanged. For a given voltage, current varies inversely with resistance. The behavior of current in a parallel circuit is best illustrated by example. The example we use is Figure 14.16. The resistors R_1, R_2, and R_3 are in parallel with each other and with the battery. Each parallel path is then a branch with its own individual current. When the total current I_T leaves the voltage source E, part I_1 of the current I_T will flow through R_1, part I_2 will flow through R_2, and the remainder I_3 through R_3. The branch current I_1, I_2, and I_3 can be different. However, if a voltmeter (used for measuring the voltage of a circuit) is connected across R_1, R_2, and R_3, the respective voltages E_1, E_2, and E_3 will be equal. Therefore,

$$E = E_1 = E_2 = E_3$$

The total current I_T is equal to the sum of all branch currents.

$$I_T = I_1 = I_2 = I_3 \tag{14.12}$$

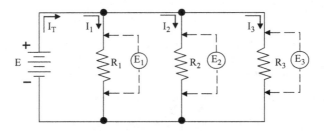

FIGURE 14.16 Parallel circuit.

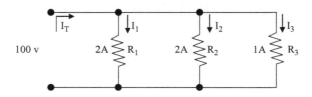

FIGURE 14.17 Refer to Example 14.18.

This formula applies for any number of parallel branches whether the resistances are equal or unequal.

By Ohm's Law, each branch current equals the applied voltage divided by the resistance between the two points where the voltage is applied. Hence, for each branch, we have the following equations (Figure 14.16):

$$\text{Branch 1: } I_1 = \frac{E_1}{R_1} = \frac{E}{R_1}$$

$$\text{Branch 2: } I_2 = \frac{E_2}{R_2} = \frac{E}{R_2} \qquad (14.13)$$

$$\text{Branch 3: } I_3 = \frac{E_3}{R_3} = \frac{E}{R_3}$$

With the same applied voltage, any branch that has less resistance allows more current through it than a branch with higher resistance.

Example 14.18

Problem:

Two resistors each drawing 2 amperes and a third resistor drawing 1 ampere are connected in parallel across a 100-V line (see Figure 14.17). What is the total current?

Solution:

The formula for total current is

$$I_T = I_1 + I_2 + I_3 \tag{14.14}$$

$$= 2 + 2 + 1$$

$$= 5 \text{ amperes}$$

The total current is 5 amperes.

Example 14.19

Problem:

Two branches R_1 and R_2 across a 100-V power line draw a total line current of 20 amperes (Figure 14.18). Branch R_1 takes 10 amperes. What is the current I_2 in branch R_2?

Solution:

Starting with equation (14.6), transpose to find I_2 and then substitute given values.

$$I_T = I_1 + I_2$$

$$I_2 = I_T - I_1$$

$$= 20 - 10 = 10 \text{ amperes}$$

The current in branch R_2 is 10 amperes.

Example 14.20

Problem:

A parallel circuit consists of two 15-Ω and one 12-Ω resistor across a 120-V line (see Figure 14.19). What current will flow in each branch of the circuit and what is the total current drawn by all the resistors?

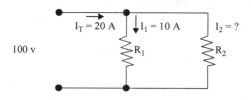

FIGURE 14.18 Refers to Example 14.19.

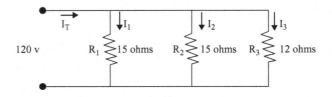

FIGURE 14.19   Refers to Example 14.20.

Solution:

There is 120-V potential across each resistor.

$$I_1 = \frac{V}{R_1} = \frac{120}{15} = 8 \text{ amperes}$$

$$I_2 = \frac{V}{R_2} = \frac{120}{15} = 8 \text{ amperes}$$

$$I_3 = \frac{V}{R_3} = \frac{120}{15} = 10 \text{ amperes}$$

Now find total current.

$$I_T = I_1 + I_2 + I_3$$
$$= 8 + 8 + 10 = 26 \text{ amperes}$$

PARALLEL RESISTANCE

Unlike series circuits where total resistance (R_T) is the sum of the individual resistances, in a parallel circuit the total resistance is *not* the sum of the individual resistances. In a parallel circuit, we can use Ohm's Law to find total resistance. We use the equation:

$$R = \frac{E}{I}$$

$$R_T = \frac{E_3}{I_t}$$

where R_T is the total resistance of all the parallel branches across the voltage source E_S, and I_T is the sum of all the branch currents.

Example 14.21

Problem:

What is the total resistance of the circuit shown in Figure 14.20?

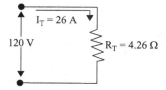

FIGURE 14.20 Refers to Example 14.21.

Given:

$$E_5 = 120 \text{ V}$$

$$I_T = 26 \text{ amperes}$$

Solution:

In Figure 14.20 the line voltage is 120 V and the total line current is 26 A. Therefore,

$$R_T = \frac{E}{I_T} = \frac{120}{26} = 4.62\,\Omega$$

Other methods used to determine the equivalent resistance of parallel circuits. The most appropriate method for a particular circuit depends on the number and value of the resistors. For this circuit, the following simple equation is used:

$$R_{eq} = \frac{R}{N} \qquad\qquad (14.15)$$

where
    R_{eq} = equivalent parallel resistance
    R = ohmic value of one resistor
    N = number of resistors
Thus,

$$R_{eq} = \frac{10\Omega}{2}$$

$$R_{eq} = 5\Omega$$

Key point: When two equal value resistors are connected in parallel, they present a total resistance equivalent to a single resistor of one-half the value of either of the original resistors.

Example 14.22

Problem:

Five 50-Ω resistors are connected in parallel. What is the equivalent circuit resistance?

Solution:

$$R_{eq} = \frac{R}{N} = \frac{50}{5} = 10\Omega$$

What about parallel circuits containing resistance of unequal value? How is equivalent resistance determined?

Example 14.23

Problem:

Given:

$$R_1 = 3\Omega$$
$$R_2 = 6\Omega$$

$E_a = 30$ V

Known:

$$I_1 = 10 \text{ amperes}$$
$$I_2 = 5 \text{ amperes}$$
$$I_t = 15 \text{ amperes}$$

Solution:

Determine:

$$R_{eq} = ?$$
$$R_{eq} = \frac{E_a}{I_t}$$
$$R_{eq} = \frac{30}{15} = 2\Omega$$

THE RECIPROCAL METHOD

When circuits are encountered in which resistors of unequal value are connected in parallel, the equivalent resistance may be computed by using the *reciprocal method*.

Note: A *reciprocal* is an inverted fraction; the reciprocal of the fraction 3/4, for example is 4/3. We consider a whole number to be a fraction with 1 as the denominator, so the reciprocal of a whole number is that number divided into 1. For example, the reciprocal of R_t is $1/R_t$. The equivalent resistance in parallel is given by the formula

$$\frac{1}{R_T} = \frac{1}{R_1} + \frac{1}{R_2} + \frac{1}{R_3} + \ldots + \frac{1}{R_n}$$ (14.16)

Where R_T is the total resistance in parallel and R_1, R_2, R_3, and R_n are the branch resistances.

Example 14.24

Problem:

Find the total resistance of a 2-Ω, a 4-Ω, and an 8-Ω resistor in parallel (Figure 14.21).

Solution:

Write the formula for three resistors in parallel.

$$\frac{1}{R_T} = \frac{1}{R_1} + \frac{1}{R_2} + \frac{1}{R_3}$$

Substituting the resistance values.

$$\frac{1}{R_T} = \frac{1}{2} + \frac{1}{4} + \frac{1}{8}$$

Add fractions.

$$\frac{1}{R_T} = \frac{4}{8} + \frac{2}{8} + \frac{1}{8} = \frac{7}{8}$$

Invert both sides of the equation to solve for R_T.

$$R_T = \frac{8}{7} = 1.14\,\Omega$$

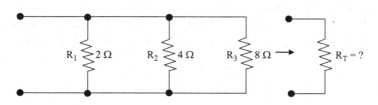

FIGURE 14.21 Refers to Example 14.24.

Note: When resistances are connected in parallel, the total resistance is always **less** than the smallest resistance of any single branch.

PRODUCT OVER THE SUM METHOD

When any two unequal resistors are in parallel, it is often easier to calculate the total resistance by multiplying the two resistances and then dividing the product by the sum of the resistances.

$$R_T = \frac{R_1 \times R_2}{R_1 + R_2} \tag{14.17}$$

Where R_T is the total resistance in parallel and R_1 and R_2 are the two resistors in parallel.

Example 14.25

Problem:

What is the equivalent resistance of a 20-Ω and a 30-Ω resistor connected in parallel?

Solution:

Given:

$$R_1 = 20$$
$$R_2 = 30$$
$$R_T = \frac{R_1 \times R_2}{R_1 + R_2}$$
$$R_T = \frac{20 \times 30}{20 + 30}$$
$$R_T = 12\Omega$$

POWER IN PARALLEL CIRCUITS

As in the series circuit, the total *power* consumed in a parallel circuit is equal to the sum of the power consumed in the individual resistors.

Note: Because power dissipation in resistors consists of a heat loss, power dissipations are additive regardless of how the resistors are connected in the circuit.

$$P_T = P_1 + P_2 + P_3 + \ldots + P_n$$

Where P_T is the total power and P_1, P_2, P_3, and P_n are the branch powers. Total power can also be calculated by the equation:

$$P_T = EI_T$$

Where P_T is the total power, E is the voltage source across all parallel branches, and I_T is the total current. The power dissipated in each branch is equal to EI and equal to V^2/R.

Note: In both parallel and series arrangements, the sum of the individual values of power dissipated in the circuit equals the total power generated by the source. The circuit arrangements cannot change the fact that all power in the circuit comes from the source.

15 Redox

ELECTROCHEMISTRY

Before presenting a discussion of battery power with lithium's contribution to producing electron flow (electricity), it is important to present information about electrochemistry. Basically and simply, this chapter is the foundation material that builds to battery power with lithium as the key player, so to speak, in electron production to drive the traction motor of an electric vehicle (EV). Battery power utilizing lithium as a major ingredient is the focus of the remainder of this text—lithium-powered batteries are currently popular for their electron flow ability.

So, let's get with it and discuss the building blocks, the processes, and the chemistry that is electrochemistry involved in the production of many of today's batteries.

In simplest terms, electrochemistry is the chemical process that causes electrons to move—remember, electricity is the movement of electrons in a circuit.

It was already pointed out what electricity is and that it is an important form of energy that we use every day in our calculators, cell phones, dishwashers, clothes washers and dryers, watches, and almost too many other applications to mention or discuss here—our focus is on lithium-battery-powered vehicles.

Again, this form of energy involves moving electrons through a wire and using the energy of these electrons. This electron movement is called electricity. Electrochemical cells used to power electronic devices and batteries. This is the case even though batteries, for example, come in a variety of different shapes and sizes, they are limited to a few basic sizes. In this chapter, some of the general information and features of various types of batteries is presented here. It is the assortment of chemical reactions involves the exchange of one or electrons—this movement of electrons is electricity. The battery is one way of producing electrical energy. The driver of electricity is called oxidation-reduction—a *redox reaction*.

ELECTROCHEMICAL REACTION—REDOX

When electrons are transferred in reactions, they are called oxidation-reduction (or "redox") reactions. The terms oxidation and reduction are used when one atom loses electrons and the other atom is absorbed. These two parts are described by the

---

FOOD FOR THOUGHT

When you get right down to it, gold, silver, diamonds, cash, and other forms of wealth, all of which can be turned to currency, there is actually only one universal currency and it is ENERGY—we all use it and we can't live without it.

DOI: 10.1201/9781003387879-19

terms "oxidation" and "reduction." It is and has been the tendency to call a substance oxidized when it is exposed to oxygen. We still use the word "oxidized" to satisfy this tendency, but today we have a much broader second meaning for this term. The loss of electrons is how we define oxidation today—it is all about losing electrons. When a substance loses electrons, its charge increases. The point is that when a substance loses electrons, its charge will increase because it has lost negative electron particles. This happening has often confused people and takes some time and thought to get used to.

Okay, keep in mind that this process is two-fold in that there is a gaining of electrons, too. Note that when an atom or an ion gains electrons, the charge on the particle actually goes down. Looking at the sulfur atom, for example, if the sulfur atom whose charge is zero received two electrons, its charge becomes (−2). Now, using another example, Fe^{3+} that gains an electron changes from +3 to +2. The bottom line is that the charge on the particle is reduced by the gain of electrons. It is important to take note and to remember that electrons have a negative charge amounting to a gain of electrons which in turn results in the charge decreasing. Now the part of all this that some find confusing and nonsensical is in *reduction*—it means gaining of electrons and the reduction of charge—and this is where the term reduction comes into play; gain of electrons but also a reduction of charge; remember, electrons have negative charge.

Note that in a chemical system oxidation and reduction have to occur concurrently and the number of electrons lost in the oxidation must be the same as the number of electrons gained in the reduction. The key point here is that whenever an oxidation-reduction reaction occurs, electrons are transferred from one substance to another. A common example of an oxidation-reduction—whenever a strip of copper is placed in a solution of silver nitrate, the following reaction takes place where the silver ions are gaining electrons to become silver atoms:

$$2Ag+(aq)+Cu(s) \rightarrow 2Ag(s)+Cu^{2+}(aq)$$

In this reaction, the copper atoms are losing electrons to become copper +2 ions, are therefore being oxidized, and the charge of copper is increasing. The point is that whenever a chemical reaction involves electrons being transferred from one substance to another, the reaction is a *redox reaction* (oxidation-reduction reaction).

Okay, the question is: Will the reaction just described produce electricity?

No ... not exactly. The reaction has to be set up in order to produce electricity. While the reaction is transferring electrons, this alone as a reaction can't produce electricity until it has a wire (conductor—the setup) for the electrons to flow through so the energy flowing can be converted to electricity. Now we can set up, the reaction, $2Ag^+$ (aq) + Cu (s) → 2Ag (s) + Cu^{2+} (aq) as one that can be arranged to produce electricity. In order to do this, the two half-reactions (oxidation and reduction) must occur in separate sections, and the separate sections remain in contact through an ionic solution (electrolyte) and external wire.

In the setup of this electrochemical cell, the copper metal must be separated from the silver ions to avoid a direct reaction. In the half reaction, each electrode in its solution could look like this:

$$Cu \rightarrow Cu^{2+} + 2e^-$$

$$2Ag^+ + 2e^- \rightarrow 2Ag$$

When a wire is connected to the two halves of the reaction, electrons flow from one metal strip to the other. In our example here. Electrons sill flow from the copper electrode (which is losing electrons) into the silver electrode (which is gaining electrons). The cell produces electricity through the wire and will continue to do so as long as there are sufficient reactants (the Ag^1 and Cu) to sustain the reaction.

When a spontaneous redox reaction occurs in an electrochemical cell, it will produce and electric current from energy released by the redox reaction.

The bottom line: It is this redox reaction that produces electric current from energy released within electrochemical cells—it is the use of lithium in these cells sets the stage for battery-powered/supplied electricity which is presented after an introduction to A-C in the following chapter. Then in Chapter 17 battery-supplied energy is explained.

16 AC Theory

INTRODUCTION[1]

So, why are we discussing alternating current (AC) theory in this chapter when we all know that electric vehicles (EVs) are driven by batteries and batteries deliver DC electricity to the prime movers (electric motors) of the EV vehicle? Well, that question is what this chapter is all about and about to answer.

First, it is magnetism, whether AC or DC is the result is not important when it comes to powering today's EVs.

Why?

Well, the truth be told, the EV can be powered by either DC or AC electrical power.

Well, an EV has a rechargeable battery—and that batteries produce DC electricity, so where does the AC power come into play?

Ah! And now we have reached the gist, that is, the point of this segment of our discussion. So, we know that an EV usually has an onboard rechargeable electrical storage unit. Generally, a lithium-ion battery that works as the source of power for an electric motor powers the EV into motion.

Motion! A key term when it comes to transportation. And motion is what travel is all about. The point being that one can't get from point A to point B to point C and beyond and then ultimately return (if so desired) without motion that is either human-accomplished physically by walking or bicycling or mechanically by driving or riding in some form of transportation—not a horse but a mechanically powered vehicle. Keep in mind that the source of power for an electric motor that powers (puts it into motion) is a battery.

Lithium-ion batteries in EVs is what powers it, the vehicle, into motion—lithium-ion batteries fit the bill, so to speak because they have high power density, high energy density, and long life as compared with others.

Okay, all the above information is great and informative but the question lingers: What motor does an EV use—is it an AC or DC motor?

As stated earlier, either AC or DC electricity is used in today's EVs. Therefore, in a science sense, it is important to have a very basic understanding of the AC just as we did with DC battery science.

BASIC AC THEORY AND PRINCIPLES

Because voltage is induced in a conductor when lines of force are cut, the amount of the induced electromotive force (emf) depends on the number of lines cut in a unit time. To induce an emf of 1 V, a conductor must cut 100,000,000 lines of force per second. To obtain this great number of "cuttings," the conductor is formed into

DOI: 10.1201/9781003387879-20

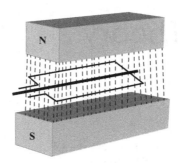

FIGURE 16.1   **Loop** rotating in magnetic field produces an AC voltage.

a loop and rotated on an axis at great speed (see Figure 16.1). The two sides of the loop become individual conductors in series, each side of the loop cutting lines of force and inducing twice the voltage that a single conductor would induce. In commercial generators, the number of "cuttings" and the resulting emf are increased by: (1) increasing the number of lines of force by using more magnets or stronger electromagnets, (2) using more conductors or loops, and (3) rotating the loops faster. [Note: both AC and DC generators are covered later.]

How an AC generator operates to produce an AC voltage and current is a basic concept today, taught in elementary and middle school science classes. Of course, we accept technological advances as commonplace, today—we surf the internet, text our friends, watch cable television, use our cell phones, and take outer space flight as a given—and consider producing the electricity that makes all these technologies possible as our right. These technologies are bottom shelf to us today—we have them available to us so we simply use them—no big deal, right? Not worth thinking about. This point of view surely was not held initially—especially by those who broke ground in developing technology and electricity.

In groundbreaking years of electric technology development, the geniuses of the science of electricity (including George Simon Ohm) performed their technological breakthroughs in faltering steps. We tend to forget that those first faltering steps of scientific achievement in the field of electricity were performed with crude, and for the most part, homemade apparatus. (Sounds something like the more contemporary young garage-and-basement inventors who came up with the first basic user-friendly microcomputer and software packages, and hopefully the renewable energy innovations of tomorrow?).

Indeed, the innovators of electricity had to fabricate nearly all the laboratory equipment used in their experiments. At the time, the only convenient source of electrical energy available to these early scientists was the voltaic cell, invented some years earlier. Because of the fact that cells and batteries were the only sources of power available, some of the early electrical devices were designed to operate from **DC**.

Thus, initially, DC was used extensively. However, when the use of electricity became widespread, certain disadvantages in the use of DC became apparent. In a DC system, the supply voltage must be generated at the level required by the load. To operate a 240-V lamp for example, the generator must deliver 240 V. A 120-V lamp could not be operated from this generator by any convenient means. A resistor could be placed in series with the 120-V lamp to drop the extra 120 V, but the resistor would waste an amount of power equal to that consumed by the lamp.

Another disadvantage of DC systems is the large amount of power loss due to the resistance of the transmission wires used to carry current from the generating station to the consumer. This loss could be greatly reduced by operating the transmission line at very high voltage and low current. This is not a practical solution in a DC system, however, since the load would also have to operate at high voltage. As a result of the difficulties encountered with DC, practically all modern power distribution systems use **AC.**

Unlike DC voltage, AC voltage can be stepped up or down by a device called a **transformer** (discussed later). Transformers permit the transmission lines to be operated at high voltage and low current for maximum efficiency. Then, at the consumer end, the voltage is stepped down to whatever value the load requires by using a transformer. Due to its inherent advantages and versatility, AC has replaced DC in all but a few commercial power distribution systems.

BASIC AC GENERATOR

As shown in Figure 16.1, an AC voltage and current can be produced when a conductor loop rotates through a magnetic field and cuts lines of force to generate an induced AC voltage across its terminals. This describes the basic principle of operation of an AC generator, or **alternator.** An alternator converts mechanical energy into electrical energy. It does this by utilizing the principle of **electromagnetic induction.** The basic components of an alternator are an armature, about which many turns of conductor are wound, which rotates in a magnetic field, and some means of delivering the resulting AC to an external circuit. [Note: We cover generator construction in more detail; in this section, we concentrate on the theory of operation].

CYCLE

An AC voltage is one that continually changes in magnitude and periodically reverses in polarity (see Figure 16.2). The zero axis is a horizontal line across the center. The vertical variations on the voltage wave show the changes in magnitude. The voltages above the horizontal axis have positive (+) polarity, while voltages below the horizontal axis have negative (–) polarity.

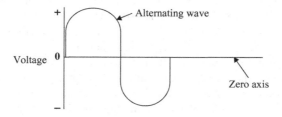

FIGURE 16.2   Basic AC voltage waveform.

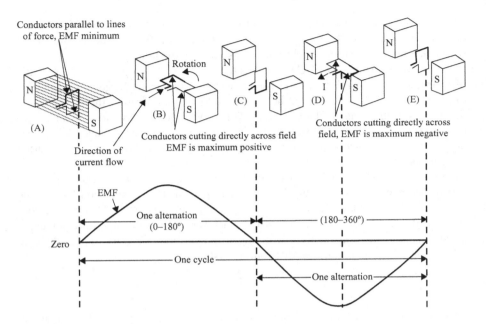

FIGURE 16.3   Basic alternating current generator.

Note: To bring the important points presented up to this point into finer focus and to expand the presentation, Figure 16.3 is provided with accompanying explanation.

Figure 16.3 shows a suspended loop of wire (conductor or armature) being rotated (moved) in a counterclockwise direction through the magnetic field between the poles of a permanent magnet. For ease of explanation, the loop has been divided into a thick and thin half. Notice that in part (A), the thick half is moving along (parallel to) the lines of force. Consequently, it is cutting none of these lines. The same is true for the thin half, moving in the opposite direction. Because the conductors are not cutting any lines of force, no emf is induced. As the loop rotates toward the position shown in part (B), it cuts more and more lines of force per second because it is cutting more directly across the field (lines of force) as it approaches the position shown in (B). At position (B), the induced voltage is greatest because the conductor is cutting directly across the field.

As the loop continues to be rotated toward the position shown in part (C), it cuts fewer and fewer lines of force per second. The induced voltage decreases from its peak value. Eventually, the loop is once again moving in a plane parallel to the magnetic field, and no voltage (zero voltage) is induced. The loop has now been rotated through half a circle (one alternation, or 180°). The sine curve shown in the lower part of Figure 16.3 shows the induced voltage at every instant of rotation of the loop. Notice that this curve contains 360°, or two alternations.

Important point: Two complete alternations in a period of time is called a *cycle*.

In Figure 16.3, if the loop is rotated at a steady rate, and if the strength of the magnetic field is uniform, the number of cycles per second (cps), or **Hertz**, and the voltage will remain at fixed values. Continuous rotation will produce a series of sine-wave voltage cycles, or, in other words, an AC voltage. In this way, mechanical energy is converted into electrical energy.

FREQUENCY, PERIOD, AND WAVELENGTH

The *frequency* of an alternating voltage or current is the number of complete cycles occurring in each second of time. It is indicated by the symbol f and is expressed in hertz (Hz). One cycle per second equals 1 Hz. Thus 60 cps equals 60 Hz. A frequency of 2 Hz (Figure 16.4b) is twice the frequency of 1 Hz (Figure 16.4a).

The amount of time for the completion of 1 cycle is the *period*. It is indicated by the symbol T for time and is expressed in seconds (s). Frequency and period are reciprocals of each other.

$$f = \frac{1}{T} \tag{16.1}$$

$$T = \frac{1}{f} \tag{16.2}$$

Important point: The higher the frequency, the shorter the period.

The angle of 360° represents the time for 1 cycle, or the period T. So we can show the horizontal axis of the sine wave in units of either electrical degrees or seconds (see Figure 16.5).

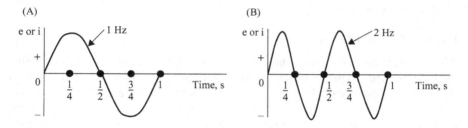

FIGURE 16.4   (a and b) Comparison of frequencies.

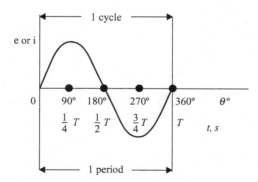

FIGURE 16.5   Relationship between electrical degrees and time.

The *wavelength* is the length of one complete wave or cycle. It depends upon the frequency of the periodic variation and its velocity of transmission. It is indicated by the symbol λ (Greek lowercase lambda). Expressed as a formula:

$$\lambda = \frac{\text{velocity}}{\text{frequency}} \tag{16.3}$$

CHARACTERISTIC VALUES OF AC VOLTAGE AND CURRENT

Because an AC sine-wave voltage or current has many instantaneous values throughout the cycle, it is convenient to specify magnitudes for comparing one wave with another. The peak, average, or root-mean-square (RMS) value can be specified (see Figure 16.6). These values apply to current or voltage.

PEAK AMPLITUDE

One of the most frequently measured characteristics of a sine wave is its amplitude. Unlike DC measurement, the amount of AC or voltage present in a circuit can be measured in various ways. In one method of measurement, the maximum amplitude of either the positive or the negative alternation is measured. The value of current or voltage obtained is called the *peak voltage* or the *peak current*. To measure the peak value of current or voltage, an oscilloscope must be used. The peak value is illustrated in Figure 16.6.

PEAK-TO-PEAK AMPLITUDE

A second method of indicating the amplitude of a sine wave consists of determining the total voltage or current between the positive and negative peaks. This value of current or voltage is called the *peak-to-peak value* (see Figure 16.6). Because both alternations of a pure sine wave are identical, the peak-to-peak value is twice the peak value. Peak-to-peak voltage is usually measured with an oscilloscope, although some voltmeters have a special scale calibrated in peak-to-peak volts.

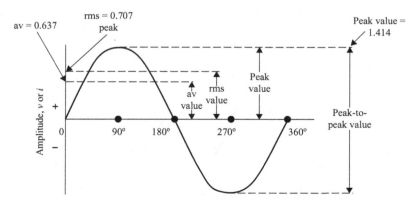

FIGURE 16.6   Amplitude values for AC sine wave.

Instantaneous Amplitude

The *instantaneous value* of a sine wave of voltage for any angle of rotation is expressed by the formula:

$$e = E_m \times \sin\theta \qquad\qquad (16.4)$$

where
  e = the instantaneous voltage
  E_m = the maximum or peak voltage
  $\sin\theta$ = the sine of angle at which e is desired

Similarly, the equation for the instantaneous value of a sine wave of current would be:

$$i = I_m \times \sin\theta \qquad\qquad (16.5)$$

where
  i = the instantaneous current
  I_m = the maximum or peak current
  $\sin\theta$ = the sine of the angle at which i desired

Note: The instantaneous value of voltage constantly changes as the armature of an alternator moves through a complete rotation. Because current varies directly with voltage, according to Ohm's Law, the instantaneous changes in current also result in a sine wave whose positive and negative peaks and intermediate values can be plotted exactly as we plotted the voltage sine wave. However, instantaneous values are not useful in solving most AC problems, so an **effective** value is used.

Effective or RMS Value

The *effective value* of an AC voltage or current of sine waveform is defined in terms of an equivalent heating effect of a DC. Heating effect is independent of the direction of current flow.

  Important point: Because all instantaneous values of induced voltage are somewhere between zero and E_M (maximum, or peak voltage), the effective value of a sine-wave voltage or current must be greater than zero and less than E_M (the maximum, or peak voltage).

  The AC of sine waveform having a maximum value of 14.14 amps produces the same amount of heat in a circuit having a resistance of one ohm as a DC of 10 amps. Because this is true, we can work out a constant value for converting any peak value to a corresponding effective value. This constant is represented by X in the simple equation below. Solve for X to three decimal places.

$$14.14X = 10$$

$$X = 0.707$$

The effective value is also called the *RMS* value because it is the square root of the average of the squared values between zero and maximum. The effective value of an AC current is stated in terms of an equivalent DC current. The phenomenon used as the standard comparison is the heating effect of the current.

Important point: Anytime an AC voltage or current is stated without any qualifications, it is assumed to be an effective value.

In many instances, it is necessary to convert from effective to peak or vice versa using a standard equation. Figure 16.6 shows that the peak value of a sine wave is 1.414 times the effective value; therefore, the equation we use is:

$$E_m = E \times 1.414 \tag{16.6}$$

where

E_m = maximum or peak voltage
E = effective or RMS voltage

and

$$I_m = I \times 1.414 \tag{16.7}$$

where

I_m = maximum or peak current
I = effective or RMS current

Upon occasion, it is necessary to convert a peak value of current or voltage to an effective value. This is accomplished by using the following equations:

$$E = E_m \times 0.707 \tag{16.8}$$

where

E = effective voltage
E_m = the maximum or peak voltage

$$I = I_m \times 0.707 \tag{16.9}$$

where

I = the effective current
I_m = the maximum or peak current

Average Value

Because the positive alternation is identical to the negative alternation, the *average value* of a complete cycle of a sine wave is zero. In certain types of circuits however, it is necessary to compute the average value of one alternation. Figure 16.6 shows that the average value of a sine wave is 0.637 × peak value and therefore:

$$\text{Average value} = 0.637 \times \text{peak value} \qquad (16.10)$$

or

$$E_{avg} = E_m \times 0.637$$

where
E_{avg} = the average voltage of one alternation
E_m = the maximum or peak voltage

Similarly,

$$I_{avg} = I_m \times 0.637 \qquad (16.11)$$

where
I_{avg} = the average current in one alternation
I_m = the maximum or peak current

Table 16.1 lists the various values of sine-wave amplitude used to multiply in the conversion of AC sine-wave voltage and current.

RESISTANCE IN AC CIRCUITS

If a sine wave of voltage is applied to a resistance, the resulting current will also be a sine wave. This follows Ohm's Law that states that the current is directly proportional to the applied voltage. Figure 16.7 shows a sine wave of voltage and the resulting sine wave of current superimposed on the same time axis. Notice that as the voltage increases in a positive direction, the current increases along with it. When the voltage reverses direction, the current reverses direction. At all times, the voltage and current pass through the same relative parts of their respective cycles at the same time. When two waves, such as those shown in Figure 16.7, are precisely in step with one another

TABLE 16.1
AC Sine Wave Conversion Table

Multiply the Value	By	To Get the Value
Peak	2	Peak-to-peak
Peak-to-peak	0.5	Peak
Peak	0.637	Average
Average	1.637	Peak
Peak	0.707	RMS (effective)
RMS (effective)	1.414	Peak
Average	1.110	RMS (effective)
RMS (effective)	0.901	Average

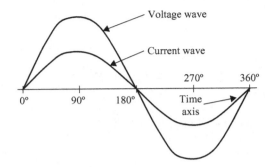

FIGURE 16.7 Voltage and current waves in phase.

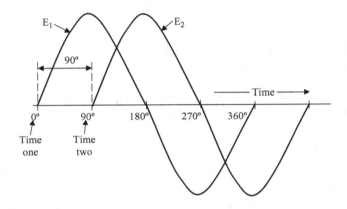

FIGURE 16.8 Voltage waves 90° out of phase.

they are said to be **in phase**. To be in phase, the two waves must go through their maximum and minimum points at the same time and in the same direction.

In some circuits, several sine waves can be in phase with each other. Thus, it is possible to have two or more voltage drops in phase with each other and also in phase with the circuit current.

Note: It is important to remember that Ohm's Law for DC circuits are applicable to AC circuits with resistance only.

Voltage waves are not always in phase. For example, Figure 16.8 shows a voltage wave E_1 considered to start at 0° (time 1). As voltage wave E_1 reaches its positive peak, a second voltage wave E_2 starts to rise (time 2). Because these waves do not go through their maximum and minimum points at the same instant of time, a **phase difference** exists between the two waves. The two waves are said to be out of phase. For the two waves in Figure 16.8, this phase difference is 90°.

PHASE RELATIONSHIPS

In the preceding section, we discussed the important concepts of **in phase** and **phase difference**. Another important phase concept is phase angle. The *phase angle* between two waveforms of the same frequency is the angular difference at a given instant of time. As an example, the phase angle between Waves B and A (see Figure 16.9) is 90°. Take the instant of time at 90°. The horizontal axis is shown in angular units of time. Wave B starts at maximum value and reduces to zero value at 90°, while Wave A starts at zero and increases to maximum value at 90°. Wave B reaches its maximum value 90° ahead of wave A, so wave B **leads** wave A by 90° (and wave A **lags** wave B by 90°). This 90° phase angle between Waves B and A is maintained throughout the complete cycle and all successive cycles. At any instant of time, Wave B has the value that Wave A will have 90° later. Wave B is a cosine wave because it is displaced 90° from Wave A, which is a sine wave.

Important point: The amount by which one wave leads or lags another is measured in degrees.

To compare phase angles or phases of alternating voltages or currents, it is more convenient to use vector diagrams corresponding to the voltage and current waveforms. A *vector* is a straight line used to denote the magnitude and direction of a given quantity. Magnitude is denoted by the length of the line drawn to scale, and the direction is indicated by the arrow at one end of the line, together with the angle that the vector makes with a horizontal reference vector.

Note: In electricity, since different directions really represent **time** expressed as a phase relationship, an electrical vector is called a **phasor**. In an AC circuit containing only resistance, the voltage and current occur at the **same time** or are in phase. To indicate this condition by means of phasors all that is necessary is to draw the phasors for the voltage and current in the same direction. The value of each is indicated by the **length** of the phasor.

A vector, or phasor, diagram is shown in Figure 16.10 where vector V_B is vertical to show the phase angle of 90° with respect to vector V_A, which is the reference. Since lead angles are shown in the counterclockwise direction from the reference vector, V_B leads V_A by 90°.

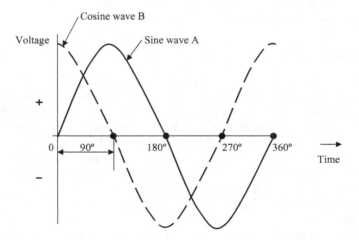

FIGURE 16.9 Wave B leads wave A by a phase angle of 90°.

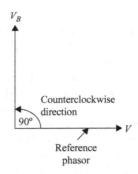

FIGURE 16.10   Phasor diagram.

INDUCTANCE

To this point, we have learned the following key points about magnetic fields:

- A field of force exists around a wire carrying a current.
- This field has the form of concentric circles around the wire, in planes perpendicular to the wire, and with the wire at the center of the circles.
- The strength of the field depends on the current. Large currents produce large fields; small currents produce small fields.
- When lines of force cut across a conductor, a voltage is induced in the conductor.

To this point, we have studied circuits that have been **resistive** (i.e., resistors presented the only opposition to current flow). Two other phenomena—inductance and capacitance –exist in DC circuits to some extent, but they are major players in AC circuits. Both inductance and capacitance present a kind of opposition to current flow that is called **reactance**, which we will cover later. Before we examine reactance, however, we must first study inductance and capacitance.

INDUCTANCE: WHAT IS IT?

Inductance is the characteristic of an electrical circuit that makes itself evident by opposing the starting, stopping, or changing of current flow. A simple analogy can be used to explain inductance. We are all familiar how difficult it is to push a heavy load (a cart full of heavy materials, etc.). It takes more work to start the load moving than it does to keep it moving. This is because the load possesses the property of **inertia**. Inertia is the characteristic of mass that opposes a **change** in velocity. Therefore, inertia can hinder us in some ways and help us in others. Inductance exhibits the same effect on current in an electric circuit as inertia does on velocity of a mechanical object. The effects of inductance are sometimes desirable—sometimes undesirable.

Important point: Simply put, **inductance** is the characteristic of an electrical conductor that opposes a **change** in current flow.

Because inductance is the property of an electric circuit that opposes any **change** in the current through that circuit, if the current increases, a self-induced voltage opposes this change and delays the increase. On the other hand, if the current decreases, a self-induced voltage tends to aid (or prolong) the current flow, delaying the decrease. Thus, current can neither increase nor decrease as fast in an inductive circuit as it can in a purely resistive circuit.

In AC circuits, this effect becomes very important because it affects the **phase** relationships between voltage and current. Earlier, we learned that voltages (or currents) can be out of phase if they are induced in separate armatures of an alternator. In that case, the voltage and current generated by each armature were in phase. When inductance is a factor in a circuit, the voltage and current generated by the **same** armature are out of phase. We shall examine these phase relationships later. Our objective now is to understand the nature and effects of inductance in an electric circuit.

Unit of Inductance

The unit for measuring inductance, L, is the *Henry* (named for the American physicist, Joseph Henry), abbreviated h and normally written in lower case, henry. Figure 16.11 shows the schematic symbol for an inductor. An inductor has an inductance of 1 henry if an emf of 1 V is induced in the inductor when the current through the inductor is changing at the rate of 1 A/s. The relation between the induced voltage, inductance, and rate of change of current with respect to time is stated mathematically as

$$E = L\frac{\Delta I}{\Delta t} \tag{16.12}$$

where
   E = the induced emf in volts
   L = the inductance in henrys
   ΔI = the change in amperes occurring in Δt seconds

Note: The symbol Δ (Delta) means "a change in ...".

The henry is a large unit of inductance and is used with relatively large inductors. The unit employed with small inductors is the millihenry (mh). For still smaller inductors, the unit of inductance is the microhenry (µh).

Self-Inductance

As previously explained, current flow in a conductor always produces a magnetic field surrounding, or linking with, the conductor. When the current changes, the

FIGURE 16.11   Schematic symbol for an inductor.

magnetic field changes, and an emf is induced in the conductor. This emf is called a *self-induced emf* because it is induced in the conductor carrying the current.

Note: Even a perfectly straight length of conductor has some inductance.

The direction of the induced emf has a definite relation to the direction in which the field that induces the emf varies. When the current in a circuit is increasing, the flux linking with the circuit is increasing. This flux cuts across the conductor and induces an emf in the conductor in such a direction to oppose the increase in current and flux. This emf is sometimes referred to as **counterelectromotive force** (cemf). The two terms are used synonymously throughout this manual. Likewise, when the current is decreasing, an emf is induced in the opposite direction and opposes the decrease in current.

Important point: The effects just described are summarized by **Lenz's Law**, which states that the induced emf in any circuit is always in a direction opposed to the effect that produced it.

Shaping a conductor so that the electromagnetic field around each portion of the conductor cuts across some other portion of the same conductor increases the inductance. This is shown in its simplest form in Figure 16.12a. A loop of conductor is looped so that two portions of the conductor lie adjacent and parallel to one another. These portions are labeled Conductor 1 and Conductor 2. When the switch is closed, electron flow through the conductor establishes a typical concentric field around **all** portions of the conductor. The field is shown in a single plane (for simplicity) that is perpendicular to both conductors. Although the field originates simultaneously in both conductors, it is considered as originating in Conductor 1 and its effect on Conductor 2 will be noted. With increasing current, the field expands outward, cutting across a portion of Conductor 2. The resultant-induced emf in Conductor 2 is shown by the dashed arrow. Note that it is in **opposition** to the battery current and voltage, according to Lenz's Law.

In Figure 16.12b, the same section of Conductor 2 is shown, but with the switch opened and the flux collapsing.

Important point: From Figure 16.12, the important point to note is that the voltage of self-induction opposes both **changes** in current. It delays the initial buildup of current by opposing the battery voltage and delays the breakdown of current by exerting an induced voltage in the same direction that the battery voltage acted.

Four major factors affect the self-inductance of a conductor, or circuit.

1. **Number of turns**—Inductance depends on the number of wire turns. Wind more turns to increase inductance. Take turns off to decrease the inductance. Figure 16.13 compares the inductance of two coils made with different numbers of turns.
2. **Spacing between turns**—Inductance depends on the spacing between turns, or the inductor's length. Figure16.14 shows two inductors with the same number of turns. The first inductor's turns have a wide spacing. The second inductor's turns are close together. The second coil, though shorter, has a larger inductance value because of its close spacing between turns.
3. **Coil diameter**— Coil diameter, or cross-sectional area, is highlighted in Figure 16.15. The larger diameter inductor has more inductance. Both coils shown have the same number of turns, and the spacing between turns is

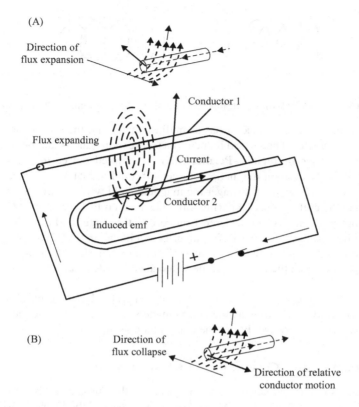

FIGURE 16.12   Self-inductance.

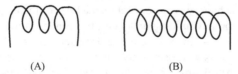

(A)   (B)

Figure 16.13   (a) Few turns, low inductance; (b) more turns, higher inductance.

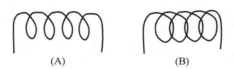

(A)   (B)

FIGURE 16.14   (a) Wide spacing between turns, low inductance; (b) close spacing between turns, higher inductance.

(A)   (B)

FIGURE 16.15   (a) Small diameter, low inductance; (b) larger diameter, higher inductance.

the same. The first inductor has a small diameter and the second one has a larger diameter. The second inductor has more inductance than the first one.

4. **Type of core material—Permeability**, as pointed out earlier, is a measure of how easily a magnetic field goes through a material. Permeability also tells us how much stronger the magnetic field will be with the material inside the coil. When comparing three identical coils, for example: One has an air core; one has a powdered-iron core in the center and the other has a soft iron core. Core material has an effect on inductance. The inductance of a coil is affected by the magnitude of current when the core is a magnetic material. When the core is air, the inductance is independent of the current.

Key point: The inductance of a coil increases very rapidly as the number of turns is increased. It also increases as the coil is made shorter, the cross-sectional area is made larger, or the permeability of the core is increased.

GROWTH & DECAY OF CURRENT IN AN RL SERIES CIRCUIT

If a battery is connected across a pure inductance, the current builds up to its final value at a rate that is determined by the battery voltage and the internal resistance of the battery. The current buildup is gradual because of the cemf generated by the self-inductance of the coil. When the current starts to flow, the magnetic lines of force move out, cut the turns of wire on the inductor, and build up a cemf that opposes the emf of the battery. This opposition causes a delay in the time it takes the current to build up to steady value. When the battery is disconnected, the lines of force collapse, again cutting the turns of the inductor and building up an emf that tends to prolong the current flow.

Although the analogy is not exact, electrical inductance is somewhat like mechanical inertia. A boat begins to move on the surface of water at the instant a constant force is applied to it. At this instant, its rate of change of speed (acceleration) is greatest, and all the applied force is used to overcome the inertia of the boat. After a while, the speed of the boat increases (its acceleration decreases) and the applied force is used up in overcoming the friction of the water against the hull. As the speed levels off and the acceleration becomes zero, the applied force equals the opposing friction force at this speed and the inertia effect disappears. In the case of inductance, it is electrical inertia that must be overcome.

MUTUAL INDUCTANCE

When the current in a conductor or coil changes, the varying flux can cut across any other conductor or coil located nearby, thus inducing voltages in both. A varying current in L_1, therefore, induces voltage across L_1 and across L_2 (see Figure 16.16; see Figure 16.17 for the schematic symbol for two coils with mutual inductance). When the induced voltage e_{L2} produces current in L_2, its varying magnetic field induces voltage in L_1. Hence, the two coils L_1 and L_2 have *mutual inductance* because current change in one coil can induce voltage in the other. The unit of mutual inductance is the henry, and the symbol is L_M. Two coils have L_M of 1 H when a current change of 1 A/s in one coil induces 1 E in the other coil.

The factors affecting the mutual inductance of two adjacent coils is dependent upon

- physical dimensions of the two coils
- number of turns in each coil
- distance between the two coils
- relative positions of the axes of the two coils
- the permeability of the cores

Important point: The amount of mutual inductance depends on the relative position of the two coils. If the coils are separated a considerable distance, the amount of flux common to both coils is small and the mutual inductance is low. Conversely, if the coils are close together so that nearly all the flow of one coil links the turns of the other mutual inductance is high. The mutual inductance can be increased greatly by mounting the coils on a common iron core.

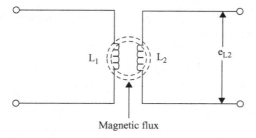

FIGURE 16.16   Mutual inductance between L1 and L2.

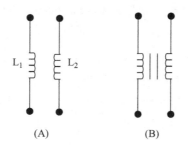

FIGURE 16.17   (a) Schematic symbol for two coils (air core) with mutual inductance; (b) two coils (iron core) with mutual inductance.

CALCULATION OF TOTAL INDUCTANCE

Note: In the study of advanced electrical theory, it is necessary to know the effect of mutual inductance in solving for total inductance in both series and parallel circuits. However, for our purposes, in this manual, we do not attempt to make these calculations. Instead, we discuss the basic total inductance calculations that the electric vehicle-on-wheels maintenance technician should be familiar with.

If inductors in series are located far enough apart, or well shielded to make the effects of mutual inductance negligible, the total inductance is calculated in the same manner as for resistances in series; we merely add them:

$$L_t = L_1 + L_2 + L_3 \dots \text{(etc.)} \qquad (16.13)$$

Example 16.1

Problem:

If a series circuit contains three inductors whose values are 40 µh, 50 µh, and 20 µh, what is the total inductance?

Solution:

$$L_t = 40 \ \mu h + 50 \ \mu h + 20 \ \mu h$$
$$= 110 \ \mu h$$

In a parallel circuit containing inductors (without mutual inductance), the total inductance is calculated in the same manner as for resistances in parallel:

$$\frac{1}{L_t} = \frac{1}{L_1} + \frac{1}{L_2} + \frac{1}{L_3} + \dots \text{(etc.)} \qquad (16.14)$$

Example 16.2

Problem:

A circuit contains three totally shielded inductors in parallel. The values of the three inductances are: 4 mh, 5 mh, and 10 mh. What is the total inductance?

Solution:

$$\frac{1}{L_t} = \frac{1}{4} + \frac{1}{5} + \frac{1}{10}$$
$$= 0.25 + 0.2 + 0.1$$
$$= 0.55$$
$$L_t = \frac{1}{0.55}$$
$$= 1.8 \ mh$$

Capacitance

No matter how complex the electrical circuit, it is composed of no more than three basic electrical properties: resistance, inductance, and capacitance. Accordingly, gaining a thorough understanding of these three basic properties is a necessary step toward the understanding of electrical equipment. Because resistance and inductance have been covered, the last of the basic three, capacitance, is covered in this section.

Earlier, we learned that inductance opposes any change in current. *Capacitance* is the property of an electric circuit that opposes any change of voltage in a circuit. That is, if applied voltage is increased, capacitance opposes the change and delays the voltage increase across the circuit. If applied voltage is decreased, capacitance tends to maintain the higher original voltage across the circuit, thus delaying the decrease.

Capacitance is also defined as that property of a circuit that enables energy to be stored in an electric field. Natural capacitance exists in many electric circuits. However, in this manual, we are concerned only with the capacitance that is designed into the circuit by means of devices called **capacitors**.

Key point: The most noticeable effect of capacitance in a circuit is that voltage can neither increase nor decrease rapidly in a capacitive circuit as it can in a circuit that does not include capacitance.

The Capacitor

A *capacitor*, or condenser, is a manufactured electrical device which consists of two conducting plates of metal separated by an insulating material called a *dielectric* (see Figure 16.18). (Note: the prefix "di-" means "through" or "across").

The schematic symbol for a capacitor is shown in Figure 16.19.

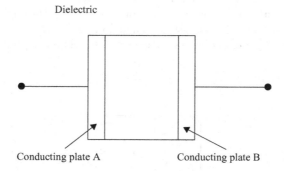

FIGURE 16.18 Capacitor.

FIGURE 16.19 (a) Schematic for a fixed capacitor; (b) variable capacitor.

When a capacitor is connected to a voltage source, there is a short current pulse. A capacitor stores this electric charge in the dielectric (it can be charged and discharged, as we shall see later). To form a capacitor of any appreciable value, however, the area of the metal pieces must be quite large and the thickness of the dielectric must be quite small.

Key point: A capacitor is essentially a device that stores electrical energy.

The capacitor is used in a number of ways in electrical circuits. It may block DC portions of a circuit since it is effectively a barrier to DC (but not to AC current). It may be part of a tuned circuit—one such application is in the tuning of a radio to a particular station. It may be used to filter AC out of a DC circuit. Most of these are advanced applications that are beyond the scope of this presentation; however, a basic understanding of capacitance is necessary to the fundamentals of AC theory.

Important point: A capacitor does not conduct DC current. The insulation between the capacitor plates blocks the flow of electrons. We learned earlier there is a short current pulse when we first connect the capacitor to a voltage source. The capacitor quickly charges to the supply voltage, and then the current stops.

The two plates of the capacitor shown in Figure 16.20 are electrically neutral since there are as many protons (positive charge) as electrons (negative charge) on each plate. Thus the capacitor has **no charge**.

Now a battery is connected across the plates (see Figure 16.21a). When the switch is closed (see Figure 16.21bB), the negative charge on plate A is attracted to the positive terminal of the battery. This movement of charges will continue until the difference in charge between plates A and B is equal to the electromotive force (voltage) of the battery.

The capacitor is now **charged**. Because almost none of the charge can cross the space between plates, the capacitor will remain in this condition even if the battery is removed (see Figure 16.22a). However, if a conductor is placed across the plates (see Figure 16.22b), the electrons find a path back to plate A and the charges on each plate are again neutralized. The capacitor is now **discharged**.

Important point: In a capacitor, electrons cannot flow through the dielectric, because it is an insulator. Because it takes a definite quantity of electrons to charge ("fill up") a capacitor, it is said to have **capacity**. This characteristic is referred to as *capacitance*.

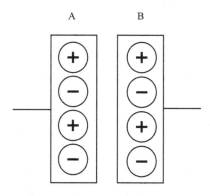

FIGURE 16.20   Two plates of a capacitor with a neutral charge.

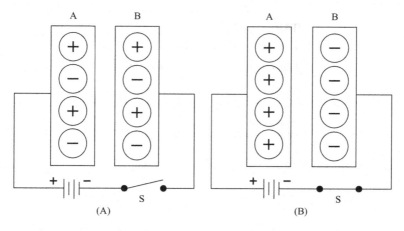

FIGURE 16.21   (a) Neutral capacitor; (b) charged capacitor.

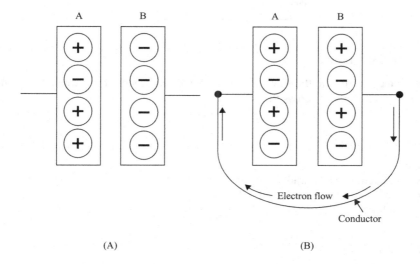

FIGURE 16.22   (a) Charged capacitor; (b) discharging a capacitor.

DIELECTRIC MATERIALS

Somewhat similar to the phenomenon of permeability in magnetic circuits, various materials differ in their ability to support electric flux (lines of force) or to serve as dielectric material for capacitors. Materials are rated in their ability to support electric flux in terms of a number called a *dielectric constant*. Other factors being equal, the higher the value of the dielectric constant, the better is the dielectric material. Dry air is the standard (the reference) by which other materials are rated.

Dielectric constants for some common materials are given in Table 16.2.

Note: From Table 16.2, it is obvious that "pure" water is the best dielectric. Keep in mind that the keyword is "pure." Water capacitors are used today in some high-energy applications, in which differences in potential are measured in thousands of volts.

TABLE 16.2
Dielectric Constants

Material	Constant
Vacuum	1.0000
Air	1.0006
Paraffin paper	3.5
Glass	5–10
Quartz	3.8
Mica	3–6
Rubber	2.5–35
Wood	2.5–8
Porcelain	5.1–5.9
Glycerine (15°C)	56
Petroleum	2
Pure water	81

UNIT OF CAPACITANCE

Capacitance is equal to the amount of charge that can be stored in a capacitor divided by the voltage applied across the plates

$$C = \frac{Q}{E} \qquad (16.15)$$

where
  C = capacitance, F (farads)
  Q = amount of charge, C (coulombs)
  E = voltage, V

Example 16.3

Problem:

What is the capacitance of two metal plats separated by one centimeter of air, if 0.002 coulomb of charge is stored when a potential of 300 V is applied to the capacitor?

Solution:

Given: Q = 0.001 coulomb

$$F = 200 \text{ V}$$

$$C = \frac{Q}{E}$$

Converting to power of ten

$$C = \frac{10 \times 10^{-4}}{2 \times 10^{2}}$$

$$C = 5 \times 10^{-6}$$

$$C = 0.000005 \text{ F}$$

Note: Although the capacitance value obtained in Example 16.3 appears small, many electronic circuits require capacitors of much smaller value. Consequently, the farad is a cumbersome unit, far too large for most applications. The **microfarad**, which is one millionth of a farad (1×10^{-6} F), is a more convenient unit. The symbols used to designated microfarad are μF.

Equation 16.15 can be rewritten as follows:

$$Q = CE \tag{16.16}$$

$$E = \frac{Q}{C} \tag{16.17}$$

Important point: From equation 16.18, do not deduce the mistaken idea that capacitance is dependent upon charge and voltage. Capacitance is determined entirely by physical factors, which are covered later.

The symbol used to designate a capacitor is (C). The unit of capacitance is the farad (F). The farad is that capacitance that will store 1 C of charge in the dielectric when the voltage applied across the capacitor terminals is 1 V.

Factors Affecting Value of Capacitance

The capacitance of a capacitor depends on three main factors: plate surface area, distance between plates, and dielectric constant of the insulating material.

- **Plate surface area**—Capacitance varies directly with place surface area. We can double the capacitance value by doubling the capacitor's plate surface area. Figure 16.23 shows a capacitor with a small surface area and another one with a large surface area.

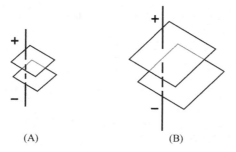

(A)   (B)

FIGURE 16.23   (a) Small plates, small capacitance; (b) larger plates, higher capacitance.

   Adding more capacitor plates can increase the plate surface area.
Figure 16.24 shows alternate plates connecting to opposite capacitor terminals.
*   **Distance between plates**—Capacitance varies inversely with the distance
    between plate surfaces. The capacitance increases when the plates are
    closer together. Figure 16.25 shows capacitors with the same plate surface
    area, but with different spacing.
*   **Dielectric constant of the insulating material**—An insulating material
    with a higher dielectric constant produces a higher capacitance rating.
    Figure 16.26 shows two capacitors. Both have the same plate surface area
    and spacing. Air is the dielectric in the first capacitor and mica is the dielec-
    tric in the second one. Mica's dielectric constant is 5.4 times greater than
    air's dielectric constant. The mica capacitor has 5.4 times more capacitance
    than the air-dielectric capacitor.

FIGURE 16.24   Several sets of plates connected to produce a capacitor with more surface area.

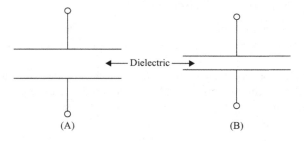

FIGURE 16.25   (a) Wide plate spacing, small capacitance; (B) narrow plate spacing, larger
capacitance.

FIGURE 16.26   (a) Low capacitance; (b) higher capacitance.

VOLTAGE RATING OF CAPACITORS

There is a limit to the voltage that may be applied across any capacitor. If too large a voltage is applied, it will overcome the resistance of the dielectric and a current will be forced through it from one plate to the other, sometimes burning a hole in the dielectric. In this event, a short circuit exists and the capacitor must be discarded. The maximum voltage that may be applied to a capacitor is known as the **working voltage** and must never be exceeded.

The working voltage of a capacitor depends on (1) the type of material used as the dielectric, and (2) the thickness of the dielectric. As a margin of safety, the capacitor should be selected so that its working voltage is at least 50% greater than the highest voltage to be applied to it. For example, if a capacitor is expected to have a maximum of 200 V applied to it, its working voltage should be at least 300 V.

CHARGE AND DISCHARGE OF AN **RC** SERIES CIRCUIT

According to Ohm's Law, the voltage across a resistance is equal to the current through it times the value of the resistance. This means that a voltage will be developed across a resistance **only when current flows through it**.

As previously stated, a capacitor is capable of storing or holding a charge of electrons. When uncharged, both plates contain the same number of free electrons. When charged, one plate contains more free electrons than the other. The difference in the number of electrons is a measure of the charge on the capacitor. The accumulation of this charge builds up a voltage across the terminals of the capacitor, and the charge continues to increase until this voltage equals the applied voltage. The greater the voltage, the greater the charge on the capacitor. Unless a discharge path is provided, a capacitor keeps its charge indefinitely. Any practical capacitor, however, has some leakage through the dielectric so that the voltage will gradually leak off.

A voltage divider containing resistance and capacitance may be connected in a circuit by means of a switch, as shown in Figure 16.27. Such a series arrangement is called an **RC series circuit**.

If S1 is closed, electrons flow counterclockwise around the circuit containing the battery, capacitor, and resistor. This flow of electrons ceases when C is charged to the battery voltage. At the instant, current begins to flow, there is no voltage on the

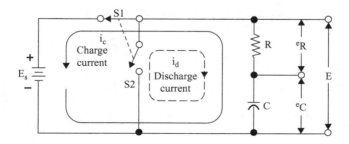

FIGURE 16.27   **Charge** and discharge of an RC series circuit.

capacitor and the drop across R is equal to the battery voltage. The initial charging current, I, is therefore equal to E_S/R.

The current flowing in the circuit soon charges the capacitor. Because the voltage on the capacitor is proportional to its charge, a voltage, e_c, will appear across the capacitor. This voltage opposes the battery voltage—that is, these two voltages buck each other. As a result, the voltage e_r across the resistor is $E_S - e_C$, and this is equal to the voltage drop $(i_c R)$ across the resistor. Because E_S is fixed, i_C decreases as e_C increases.

The charging process continues until the capacitor is fully charged and the voltage across it is equal to the battery voltage. At this instant, the voltage across is zero and no current flows through it.

If S2 is closed (S1 opened) in Figure 16.27, a discharge current, i_d, will discharge the capacitor. Because i_t is opposite in direction to i_c, the voltage across the resistor will have a polarity opposite to the polarity during the charging time. However, this voltage will have the same magnitude and will vary in the same manner. During discharge, the voltage across the capacitor is equal and opposite to the drop across the resistor. The voltage drops rapidly from its initial value and then approaches zero slowly.

The actual time it takes to charge or discharge is important in advanced electricity and electronics. Because the charge or discharge time depends on the values of resistance and capacitance, an RC circuit can be designed for the proper timing of certain electrical events. RC time constant is covered in the next section.

RC TIME CONSTANT

The time required to charge a capacitor to 63% of maximum voltage or to discharge it to 37% of its final voltage is known as the *time constant* of the current. An RC circuit is shown in Figure 16.28.

The time constant T for a RC circuit is

$$T = RC \tag{16.18}$$

The time constant of an RC circuit is usually very short because the capacitance of a circuit may be only a few microfarads or even picofarads.

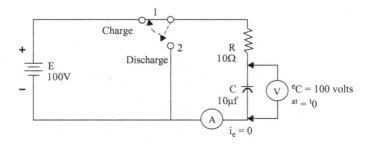

FIGURE 16.28   RC circuit.

Key point: An RC time constant expresses the charge and discharge times for a capacitor.

CAPACITORS IN SERIES AND PARALLEL

Like resistors or inductors, capacitors may be connected in series, in parallel, or in a series-parallel combination. Unlike resistors or inductors, however, total capacitance in series, in parallel, or in a series-parallel combination is found in a different manner. Simply put, the rules are not the same for the calculation of total capacitance. This difference is explained as follows:

Parallel capacitance is calculated like series resistance, and series capacitance is calculated like parallel resistance. For example:

When capacitors are connected in *series* (see Figure 16.29), the total capacitance C_T is

$$\text{Series: } \frac{1}{C_T} = \frac{1}{C_1} + \frac{1}{C_2} + \frac{1}{C_3} + \ldots + \frac{1}{C_n} \tag{16.19}$$

Example 16.4

Problem:

Find the total capacitance of a 3-μF, a 5-μF, and a 15-μF capacitor in series.

Solution:

Write equation 16.20 for three capacitors in series.

$$\frac{1}{C_T} = \frac{1}{C_1} + \frac{1}{C_2} + \frac{1}{C_3}$$

$$\frac{1}{3} + \frac{1}{5} + \frac{1}{15} = \frac{9}{15} = \frac{3}{5} = \frac{5}{3} = 1.7\,\mu F$$

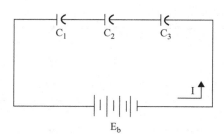

FIGURE 16.29 Series capacitive circuit.

When capacitors are connected in **parallel** (see Figure 16.30), the total capacitance C_T is the sum of the individual capacitances.

$$\text{Parallel: } C_T = C_1 + C_2 + C_3 + ... + C_n \qquad (16.20)$$

Example 16.5

Problem:

Determine the total capacitance in a parallel capacitive circuit:

Given: $C_1 = 2\ \mu F$

$C_2 = 3\ \mu F$

$C_3 = 0.25\ \mu F$

Solution:

Write equation 16.22 for three capacitors in parallel:

$$C_T = C_1 + C_2 + C_3$$
$$= 2 + 3 + 0.25$$
$$= 5.25\ \mu F$$

Capacitors can be connected in a combination of *series and parallel* (see Figure 16.31).

Types of Capacitors

Capacitors used for commercial applications are divided into two major groups—fixed and variable—and are named according to their dielectric. Most common are air, mica, paper, and ceramic capacitors, plus the electrolytic type. These types are compared in Table 16.3.

The fixed capacitor has a set value of capacitances that is determined by its construction. The construction of the variable capacitor allows a range of capacitances. Within this range, the desired value of capacitance is obtained by some mechanical

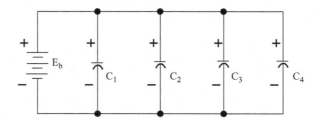

FIGURE 16.30   Parallel capacitive circuit.

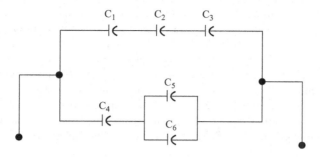

FIGURE16.31   Series-parallel capacitance configuration.

TABLE 16.3
Comparison of Capacitor Types

Dielectric	Construction	Capacitance Range
Air	Meshed plates	10–400 pF
Mica	Stacked plates	10–5,000 pF
Paper	Rolled foil	0.001-1 µF
Ceramic	Tubular	0.5–1,600 pF
	Disk	0.002-0.1 µF
Electrolytic	Aluminum	5–1,000 µF
	Tantalum	0.01–300 µF

means, such as by turning a shaft (as in turning a radio tuner knob, for example) or adjusting a screw to adjust the distance between the plates.

The electrolytic capacitor consists of two metal plates separated by an electrolyte. The electrolyte, either paste or liquid, is in contact with the negative terminal, and this combination forms the negative electrode. The dielectric is a very thin film of oxide deposited on the positive electrode, which is aluminum sheet. Electrolytic capacitors are polarity sensitive (i.e., they must be connected in a circuit according to their polarity markings) used where a large amount of capacitance is required.

ULTRACAPACITORS

NREL (2009) points out that like batteries, ultracapacitors, also known as supercapacitors, pseudocapacitors, electric double-layer capacitors, or electrochemical double layer capacitors (EDLCs), are energy storage devices. To meet the power, energy, and voltage requirements for a wide range of applications, they use electrolytes and configure various-sized cells into modules. As storage devices, however, they differ from batteries in that ultracapacitors (which are true capacitors in that energy is stored via charge separation at the electrode-electrolyte interface) store energy electrostatically, whereas batteries store energy chemically. Not only do ultracapacitors provide quick bursts of energy, they also are an improvement in about two to three orders of magnitude in capacitance (as compared to an average capacitor), but with a lower working voltage. Moreover, they can withstand hundreds of thousands or charge/discharge cycles without degrading. As an alternative energy source, ultracapacitors have proven themselves as reliable energy storage components used to power a variety of electronic and portable devices such as AM/FM radios, flashlights, cell phones, and emergency kits. As ultracapacitor technology matures they are being developed to function as batteries; for example, the vehicle industry is deploying ultracapacitors as a replacement for chemical batteries. In the automotive industry, they have been integrated into EV and hybrid electric vehicles (HEV) to help alleviate stress and extend the life of the batteries.

Ultracapacitor Operation

An ultracapacitor polarizes an electrolytic solution to store energy electrostatically. Though it is an electrochemical device, no chemical reactions are involved in its energy storage mechanism. This mechanism is highly reversible and allows the ultracapacitor to be charged and discharged hundreds of thousands of times.

An ultracapacitor can be viewed as two nonreactive porous plates, or collectors, suspended within an electrolyte, with a voltage potential applied across the collectors. In an individual ultracapacitor cell, the applied potential on the positive electrode attracts the negative ions in the electrolyte, while the potential on the negative electrode attracts the positive ions. A dielectric separator between the two electrodes prevents the charge from moving between the two electrodes. Once the ultracapacitor is charged and energy stored, a load can use this energy. The amount of energy stored is very large compared to a standard capacitor because of the enormous surface area created by the porous carbon electrodes and the small separation (10 Å) created by the dielectric separate. However, it stores a much smaller amount of energy than does a battery. Because the rates of charge and discharge are determined solely by is physical properties, the ultracapacitor can release energy much faster (with more power) than a battery that relies on slow chemical reactions.

INDUCTIVE & CAPACITIVE REACTANCE

Earlier, we learned that the inductance of a circuit acts to oppose any change of current flow in that circuit and that capacitance acts to oppose any change of voltage. In DC circuits these **reactions** are not important, because they are momentary and occur only when a circuit is first closed or opened. In AC circuits, these effects become very important because the direction of current flow is reversed many times each second; and the opposition presented by inductance and capacitance is, for practical purposes, constant.

In purely resistive circuits, either DC or AC, the term for opposition to current flow is resistance. When the effects of capacitance or inductance are present, as they often are in AC circuits, the opposition to current flow is called *reactance*. The total opposition to current flow in circuits that have both resistance and reactance is called *impedance*.

In this section, we cover the calculation of inductive and capacitive reactance and impedance; the phase relationships of resistance, inductive, and capacitive circuits; and power in reactive circuits.

INDUCTIVE REACTANCE

In order to gain understanding of the reactance of a typical coil, we need to review exactly what occurs when AC voltage is impressed across the coil.

1. The AC voltage produces an AC.
2. When a current flows in a wire, lines of force are produced around the wire.
3. Large currents produce many lines of force; small currents produce only a few lines of force.
4. As the current changes, the number of lines of force will change. The field of force will seem to expand and contract as the current increases and decreases as shown in Figure 16.32.
5. As the field expands and contracts, the lines of force must cut across the wires that form the turns of the coil.
6. These cuttings induce an emf in the coil.
7. This emf acts in the direction so as to oppose the original voltage and is called a *counter*, or back, *emf.*
8. The effect of this cemf is to reduce the original voltage impressed on the coil. The net effect will be to reduce the current below that which would flow if there were no cuttings or cemf.
9. In this sense, the cemf is acting as a resistance in reducing the current.
10. Although it would be more convenient to consider the current reducing effect of a cemf as a number of ohms of effective resistance, we don't do this. Instead, since a cemf is not actually a resistance but merely **acts** as a resistance, we use the term *reactance* to describe this effect.

Important point: The **reactance** of a coil is the number of ohms of resistance that the coil **seems** to offer as a result of a cemf induced in it. Its symbol is X to differentiate it from the DC resistance R.

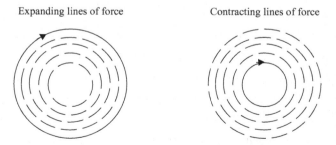

FIGURE 16.32   An AC current producing a moving (expanding and collapsing) field. In a coil, this moving field cuts the wires of the coil.

The inductive reactance of a coil depends primarily on (1) the coil's inductance and (2) the frequency of the current flowing through the coil. The value of the reactance of a coil is therefore proportional to its inductance and the frequency of the AC circuit in which it is used.

The formula for inductive reactance is

$$X_L = 2\pi fL \tag{16.21}$$

Since $2\pi = 2(3.14) = 6.28$, equation 16.22 becomes

$$X_L = 6.28\,fL$$

where
X_L = inductive reactance, Ω
f = frequency, Hz
L = inductance, H

If any two quantities are known in equation 16.21, the third can be found.

$$L = \frac{X_L}{6.28\,f} \tag{16.22}$$

$$f = \frac{X_L}{6.28L} \tag{16.23}$$

Example 16.6

Problem:

The frequency of a circuit is 60 Hz and the inductance is 20 mh. What is X_L?

Solution:

$$X_L = 2\pi fL$$
$$= 6.28 \times 60 \times 0.02$$
$$= 7.5\,\Omega$$

Example 16.7

Problem:

A 30-mh coil is in a circuit operating at a frequency of 1,400 kHz. Find its inductive reactance.

Solution:

Given: $L = 30$ mh

    $f = 1400$ kHz

Find $= X_L = ?$

Step 1: Change units of measurement.

    30 mH $= 30 \times 10^{-3}$ h

    1,400 kHz $= 1,400 \times 10^{3}$

Step 2: Find the inductive reactance.

$$X_L = 6.28\,fL$$
$$X_L = 6.28 \times 1,400 \times 10^{3} \times 30 \times 10^{-3}$$
$$X_L = 263,760\ \Omega$$

Example 16.8

Problem:

Given: $L = 400$ µh

    $f = 1,500$ Hz

    Find $X_L = ?$

Solution:

$$X_L = 2\pi fL$$
$$= 6.28 \times 1,500 \times 0.0004$$
$$= 3.78\ \Omega$$

Key point: If frequency or inductance varies, inductive reactance must also vary. A coil's inductance does not vary appreciably after the coil is manufactured unless it is designed as a variable inductor. Thus, frequency is generally the only variable factor affecting the inductive reactance of a coil. The coil's inductive reactance will vary directly with the applied frequency.

CAPACITIVE REACTANCE

Previously, we learned that as a capacitor is charged, electrons are drawn from one plate and deposited on the other. As more and more electrons accumulate on the second plate, they begin to act as an opposing voltage which attempts to stop the flow of electrons just as a resistor would do. This opposing effect is called the *reactance* of the capacitor and is measured in ohms. The basic symbol for reactance is X, and the subscript defines the type of reactance. In the symbol for inductive reactance,

X_L, the subscript L refers to inductance. Following the same pattern, the symbol for capacitive reactance is X_C.

Key point: **Capacitive reactance, X_C,** is the opposition to the flow of AC current due to the capacitance in the circuit.

The factors affecting capacitive reactance, X_C, are:

*   the size of the capacitor
*   frequency.

The larger the capacitor, the greater the number of electrons that may be accumulated on its plates. However, because the plate area is large, the electrons do not accumulate in one spot but spread out over the entire area of the plate and do not impede the flow of new electrons onto the plate. Therefore, a large capacitor offers a small reactance. If the capacitance were small, as in a capacitor with a small plate area, the electrons could not spread out and would attempt to stop the flow of electrons coming to the plate. Therefore, a small capacitor offers a large reactance. The reactance is therefore **inversely** proportional to the capacitance.

If an AC voltage is impressed across the capacitor, electrons are accumulated first on one plate and then on the other. If the frequency of the changes in polarity is low, the time available to accumulate electrons will be large. This means that a large number of electrons will be able to accumulate, which will result in a large opposing effect, or a large reactance. If the frequency is high, the time available to accumulate electrons will be small. This means that there will be only a few electrons on the plates, which will result in a small opposing effect, or a small reactance. The reactance is, therefore, **inversely** proportional to the frequency.

The formula for capacitive reactance is

$$X_C = \frac{1}{2\pi f C} \tag{16.24}$$

with C measured in farads.

Example 16.9

Problem:

What is the capacitive reactance of a circuit operating at a frequency of 60 Hz, if the total capacitance is 130 μF?

Solution:

$$X_C = \frac{1}{2\pi f C}$$

$$= \frac{1}{6.28 \times 60 \times 0.00013}$$

$$= 20.4 \ \Omega$$

Phase Relationship R, L, & C Circuits

Unlike a purely resistive circuit (where current rises and falls with the voltage; that is, it neither leads nor lags and current and voltage are in phase), current and voltage are not in phase in inductive and capacitive circuits. This is the case, of course, because occurrences are not quite instantaneous in circuits that have either inductive or capacitive components.

In the case of an inductor, voltage is first applied to the circuit, then the magnetic field begins to expand, and self-induction causes a counter current to flow in the circuit, opposing the original circuit current. Voltage **leads** current by 90° (see Figure 16.33).

When a circuit includes a capacitor, a charge current begins to flow and then a difference in potential appears between the plates of the capacitor. Current **leads** voltage by 90° (see Figure 16.34).

Key point: In an inductive circuit, voltage leads current by 90°; and in a capacitive circuit, current leads voltage by 90°.

Impedance

Impedance is the total opposition to the flow of AC in a circuit that contains resistance and reactance. In the case of pure inductance, inductive reactance, X_L is the

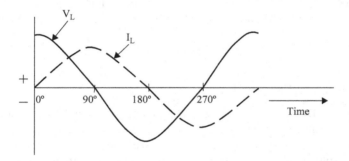

FIGURE 16.33   Inductive circuit—voltage leads current by 90°.

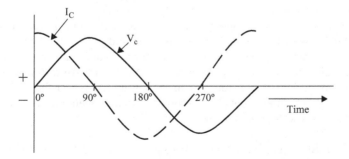

FIGURE 16.34   Capacitive circuit—current leads voltage by 90°.

total opposition to the flow of current through it. In the case of pure resistance, R represents the total opposition. The combined opposition of R and X_L in series or in parallel to current flow is called **impedance**. The symbol for impedance is Z.

The impedance of resistance in series with inductance is

$$Z = \sqrt{R^2 + X_L^2} \tag{16.25}$$

where
Z = impedance, Ω
R = resistance, Ω
X_L = inductive reactance, Ω

The impedance of resistance in series with capacitance is:

$$Z = \sqrt{R^2 + X_C^2} \tag{16.26}$$

where
Z = impedance, Ω
R = resistance, Ω
X_C = inductive capacitance, Ω

When the impedance of a circuit includes R, X_L, and X_C, both resistance and net reactance must be considered. The equation for impedance, including both X_L and X_C is:

$$Z = \sqrt{R^2 + (X_L - X_C)^2} \tag{16.27}$$

POWER IN REACTIVE CIRCUITS

The power in a DC circuit is equal to the product of volts and amps, but in an AC circuit, this is true only when the load is resistive and has no reactance.

In a circuit possessing inductance only, the true power is zero. The current lags the applied voltage by 90°. The true power in a capacitive circuit is also zero. The *true power* is the average power actually consumed by the circuit; the average being taken over one complete cycle of AC. The *apparent power* is the product of the RMS volts and rms amps.

The ratio of true power to apparent power in an AC circuit is called the *power factor*. It may be expressed as a percent or as a decimal.

To this point, we have pointed out how a combination of inductance and resistance and then capacitance and resistance behave in an AC circuit. We saw how the RL and RC combination affects the current, voltages, power, and power factor of a circuit. We considered these fundamental properties as isolated phenomena. The following phase relationships were seen to be true:

1. The voltage drop across a resistor is **in phase** with the current through it.
2. The voltage drop across an inductor **leads** the current through it by 90°.

3. The voltage drop across a capacitor **lags** the current through it by 90°.
4. The voltage drops across inductors and capacitors are **180 degrees out of phase.**

Solving AC problems is complicated by the fact that current varies with time as the AC output of an alternator goes through a complete cycle. This is the case because the various voltage drops in the circuit vary in-phase—they are not at their maximum or minimum values at the same time.

AC circuits frequently include all three circuit elements: resistance, inductance, and capacitance. In this section, all three of these fundamental circuit parameters are combined and their effect on circuit values studied.

NOTE

1 This chapter is based on F.R. Spellman's (2023) *The Science of Electric Vehicles: Concepts and Applications.* Boca Raton FL: CRC Press.

REFERENCES

NREL (2009). *Ultracapacitors.* National Renewable Energy Laboratory. Accessed 01/29/10 @ http://www.nrel.gov/vehiclesandfuels/energystorage/ultracapacitors.html?print.

17 Battery-Supplied Energy

THE 411 ON BATTERIES

Simply, when talking about lithium-powered batteries, it is a subject that is gaining in popularity and use. With regard to batteries, they are used universally for several applications. Keep in mind that the energy stored in the battery provides direct current (DC) electricity—again, with many widespread applications and is widely used in household, commercial, and industrial operations—also commonly used for vehicle lighting and ignition of vehicles. Beyond their applications in vehicles, other somewhat well-known applications of battery-supplied systems include providing electrical energy in industrial vehicles and emergency diesel generators, material handling equipment (forklifts), portable electric/electronic equipment, backup emergency power for light-packs, for hazard warning signal lights and flashlights, and as standby power supplies or uninterruptible power supplies (UPS) for computer systems. In some instances, they are used as the only source of power; while in others (as mentioned above), they are used as a secondary or standby power supply. In renewable energy applications, batteries are used to store electrical energy. A battery pack stores electricity produced by a solar electric system. Today, batteries commonly are used in specialized applications such as plug-in *battery electric vehicle (BEV) and hybrid electric vehicle (HEV)* rechargeable energy storage systems. Batteries are often used in wind-hybrid systems to store excess wind energy and then provide supplementary energy when the wind cannot generate sufficient power to meet the electric load. Batteries are also used and in road side solar-charging systems, solar energy is used to charge a battery pack that supplies electrical power to some type of electrical signaling device, navigation buoys, or other remote application (e.g., emergency telephone) where it would be impractical and costly to run miles of electrical supply cable.

Note that in this book our main focus is on lithium batteries used in **electric vehicle battery** (**EVB** also known as **a traction battery**) systems that are rechargeable and used to power the electric motors of BEVs or HEVs. These EVB's are typically **lithium-ion batteries** (**LIB**); they are specifically designed for high electric charge (aka energy) capacity. EVBs are discussed in detail in this chapter; however, this is a science book so to properly introduce any type of battery, it is important to begin with battery basics and then advancing, step-by-step, into a discussion of the presently available EVBs and their use of lithium batteries.

BASIC BATTERY TERMINOLOGY

- A voltaic cell is a combination of materials used to convert chemical energy into electrical energy.
- A battery is a group of two or more connected voltaic/galvanic cells connected in series, where the battery potential equals the cells' potentials.

DOI: 10.1201/9781003387879-21

- An electrode is a metallic compound, or metal, which has an abundance of electrons (negative electrode) or an abundance of positive charges (positive electrode).
- An electrolyte is a solution which is capable of conducting an electric current.
- Specific gravity is defined as the ratio comparing the weight of any liquid to the weight of an equal volume of water.
- An ampere-hour is defined as a current of one ampere flowing for 1 hour.
- An external circuit is used to conduct the flow of electrons between the electrodes of the voltaic cell and usually includes a load.
- Three types of batteries:

  - Primary batteries are not rechargeable.
  - Secondary batteries are rechargeable.
  - Fuel cells need a supply of fuel to provide the energy.

DID YOU KNOW?

Primary batteries include dry, alkaline, and mercury batteries. A new dry cell has a potential of about 1.5 V regardless of size, but the energy a battery can deliver depends on its size.

SECONDARY BATTERY CHARACTERISTICS

Secondary batteries (commonly called *rechargeable batteries*) are designed to be depleted of energy and then recharged and used again and again as opposed to the primary batteries that can be discharged once and then needs to be disposed of.

Secondary batteries are generally classified by their various characteristics, some of which can be used to classify other types of batteries. Parameters such as internal resistance, cycle life, cell voltage, capacity, specific energy, shelf life, operating life, self-discharge, specific gravity, "c" rate, mid-point voltage (MPV), gravimetric energy density, volumetric energy density, constant-voltage charges, constant-current charge, and specific power are used to describe and classify batteries by operation and type.

Regarding *internal resistance*, it is important to keep in mind that a battery is a DC voltage generator. As such, the battery has internal resistance or equivalent series resistance (ESR). In a chemical cell, the resistance of the electrolyte between the electrodes is responsible for most of the cell's internal resistance. Because any current in the battery must flow through the internal resistance, this resistance is in series with the generated voltage. With no current, the voltage drop across the resistance is zero so that the full generated voltage develops across the output terminals. This is the open-circuit voltage, or no-load voltage. If a load resistance is connected across the battery, the load resistance is in series with internal resistance. When current flows in this circuit, the internal voltage drop decreases the terminal voltage of the battery.

How many times a battery can be discharged and recharged is known as *cycle life*.

The voltage supplied by a single battery cell and measured in four different ways—float, nominal, charge, and discharge—is called *cell voltage*.

The *capacity* of a battery is measured in ampere-hours (Ah), milli Ampere hours (mAh), or cold cranking amps; these parameters define how long a battery can power its electrical load (appliances, etc.). The ampere-hour capacity is equal to the product of the current in amperes and the time in hours during which the battery is supplying this current and is defined as the amount of current that a battery can deliver for 1 hour before the battery voltage reaches the end-of-life point. In other words, 1 Ah is the equivalent of drawing 1 amp steadily for 1 hour, or 2 amps steadily for half an hour. A typical 12-V system may have 800 Ah of battery capacity. The battery can draw 100 amps for 8 hours if fully discharged and starting from a fully charged state. This is equivalent to 1,200 watts for 8 hours (power in watts = amps × volts). The ampere-hour capacity varies inversely with the discharge current. The size of a cell is determined generally by its ampere-hour capacity.

The capacity of a storage battery determines how long it will operate at a given discharge rate and depends upon many factors, the most important of these are as follows:

- The area of the plates in contact with the electrolyte
- The quantity and specific gravity of the electrolyte
- The type of separators
- The general condition of the battery (degree of sulfating, plates bucked, separators warped, sediment in bottom of cells, etc.)
- The final limiting voltage

DID YOU KNOW?

For the average person who hears or reads the term, ampere hour, usually has difficulty in understanding what the term actually means. The easy to learn and understand the meaning of ampere hour is to think of how long it will last while connected to its load.

The *specific energy* of a battery is the number of Watt hours of energy that can be supplied by a 1-kilogram battery.

The *shelf life* of a cell is that period of time during which the cell can be stored without losing more than approximately 10% of its original capacity. The loss of capacity of a stored cell is due primarily to the drying out of its electrolyte in a wet cell and to chemical actions which change the materials within the cell. The shelf life of a cell can be extended by keeping it in a cool, dry place.

The life of a battery or more commonly known as the *operating life* of the battery is just what the name implies—how long the battery will last in units of time (e.g., years) if maintained and used often.

A percentage over times such as 8% annually is known as the *self-discharge*—how much energy the battery loses when not in use and not on charge

The ratio of the weight of a certain volume of liquid to the weight of the same volume of water is called the *specific gravity* of the liquid. Pure sulfuric acid has a specific gravity of 1.835 since it weighs 1.835 times as much as water per unit volume. The specific gravity of a mixture of sulfuric acid and water varies with the strength of the solution from 1.000 to 1.830.

The specific gravity of the electrolyte solution in a lead-acid cell ranges from 1.210 to 1.300 for new, fully charged batteries. The higher the specific gravity, the less internal resistance of the cell and the higher the possible load current. As the cell discharges, the water formed dilutes the acid and the specific gravity gradually decreases to about 1.150, at which time the cell is considered to be fully discharged.

The specific gravity of the electrolyte is measured with a *hydrometer*, which has a compressible rubber bulb at the top, a glass barrel, and a rubber hose at the bottom of the barrel. In taking readings with a hydrometer, the decimal point is usually omitted. For example, a specific gravity of 1.260 is read simply as "twelve-sixty." A hydrometer reading of 1,210–1,300 indicates full charge; about 1,250 is half-charge; and 1,150–1,200 is complete discharge.

The *"c" rate* is a current that is numerically equal to the Ah-rating of the cell. Charge and discharge currents are typically expressed in fractions of multiples of the c rate. Stated as simply as possible c rate stands for the capacity of a battery usually measured in ampere-hours (Ah) indicating the amount of active material with the battery available for discharge. A cell having a capacity of 20-Ah, for example, should deliver 20 amperes of electron flow (current) for 1 hour or 2 amperes (aka amps) of current for 10 hours.

The **MPV** is the nominal voltage of the cell and is the voltage that is measured when the battery has discharged 50% of its total energy.

The **gravimetric energy density** of a battery is a measure of how much energy a battery contains in comparison to its weight.

The **volumetric energy density** of a battery is a measure of how much energy a battery contains in comparison to is volume.

A **constant-voltage charger** is a circuit that recharges a battery by sourcing only enough current to force the battery voltage to a fixed value.

A **constant-current charger** is a circuit that charges a battery by sourcing a fixed current into the battery, regulates of battery voltage.

In batteries, **specific power** usually refers to the power-to-weight ratio, measured in kilowatts per kilogram (generally, kW/kg).

DID YOU KNOW?

Note that the Amp hour rating for a battery, of whatever type, is only as accurate what you know about the battery. That is, what was the temperature of area in which the battery was tested in. Then it is also important to know the hour rate that was used. Finally, when the measurement of Amp hour was recorded what was the battery powering?

Common secondary battery chemistries include *nickel* (as in nickel cadmium, nickel metal hydride, nickel iron, nickel zinc, and nickel hydrogen); lithium ion (as in lithium-ion cobalt oxide, lithium-ion manganese oxide, lithium-ion iron phosphate, lithium-ion nickel cobalt aluminum oxide, and so forth; and lead acid (e.g., sealed and flooded lead-acid batteries).

A ***lithium-ion battery*** is a rechargeable battery with an anode made up of lithium compounds. LIB have a chip inside them to help manage internal processes. These complex batteries should never be completely discharged. Complete discharge, deep discharge, can be harmful to the battery. This is why at the 80% discharge rate it is recommended that the application being used be stopped in order to recharge the battery.

Note that lithium batteries are in high demand in commercial activities because of their assorted features. The market for the increased usage of lithium batteries is predicted to expand exponentially as more and more applications are powered by the lithium cells. And a huge benefit of the lithium ions' lithium compound chemistry, LIB can be recharged 100s of times—this is not the case.

Okay, now it is time for us to delve into basic battery construction and operation; we cover the basics in the following section.

VOLTAIC CELL

The simplest cell (an electrochemical device that transforms chemical energy into electrical energy) is known as a *voltaic* (or galvanic) cell (see Figure 17.1). It consists of a piece of carbon (C) and a piece of zinc (Zn) suspended in a jar that contains a solution of water (H_2O) and sulfuric acid (H_2SO_4).

Note: A simple cell consists of two strips, or **electrodes**, placed in a container that holds the **electrolyte**. A battery is formed when two or more cells are connected.

The electrodes are the conductors by which the current leaves or returns to the electrolyte. In the simple cell described above, they are carbon and zinc strips placed in the electrolyte. Zinc contains an abundance of negatively charged atoms, while carbon has an abundance of positively charged atoms. When the plates of these materials are immersed in an electrolyte, chemical action between the two begins.

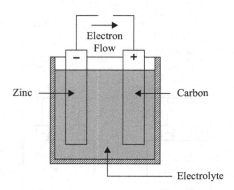

FIGURE 17.1 Simple voltaic cell.

In the **dry cell** (see Figure 17.2), the electrodes are the carbon rod in the center and the zinc container in which the cell is assembled.

The electrolyte is the solution that acts upon the electrodes that are placed in it. The electrolyte may be a salt, an acid, or an alkaline solution. In the simple voltaic cell and in the automobile storage battery, the electrolyte is in a liquid form; while in the dry cell (see Figure 17.2), the electrolyte is a moist paste.

PRIMARY AND SECONDARY CELLS

Primary cells are normally those that cannot be recharged or returned to good condition after their voltage drops too low. Dry cells in flashlights and transistor radios are examples of primary cells. Some primary cells have been developed to the state where they can be recharged.

A *secondary cell* is one in which the electrodes and the electrolyte are altered by the chemical action that takes place when the cell delivers current. These cells are rechargeable. During recharging, the chemicals that provide electric energy are restored to their original condition. Recharging is accomplished by forcing an electric current through them in the opposite direction to that of discharge.

Connecting as shown in Figure 17.3 recharges a cell. Some battery chargers have a voltmeter and an ammeter that indicate the charging voltage and current.

The automobile storage battery is the most common example of the secondary cell.

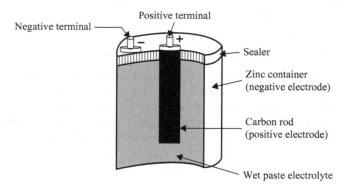

FIGURE 17.2   Dry cell (cross-sectional view).

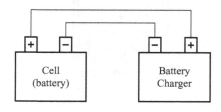

FIGURE 17.3   Hookup for charging a secondary cell with a battery charger.

BATTERY

As stated previously, a cell is an electrochemical device capable of supplying the energy that results from an internal chemical reaction to an external electrical circuit. In the simplest terms, batteries are made up of an anode, cathode, separate, electrolyte, and two current collectors (positive and negative). A *battery* consists of two or more cells placed in a common container. The cells are connected in series, in parallel, or in some combination of series and parallel, depending upon the amount of voltage and current required of the battery. The connection of cells in a battery is discussed in more detail later.

BATTERY OPERATION

The chemical reaction within a battery provides the voltage. This occurs when a conductor is connected externally to the electrodes of a cell, causing electrons to flow under the influence of a difference in potential across the electrodes from the zinc (negative) through the external conductor to the carbon (positive), returning within the solution to the zinc. After a short period of time, the zinc will begin to waste away because of the acid.

The voltage across the electrodes depends upon the materials from which the electrodes are made and the composition of the solution. The difference of potential between the carbon and zinc electrodes in a dilute solution of sulfuric acid and water is about 1.5 V.

The current that a primary cell may deliver depends upon the resistance of the entire circuit, including that of the cell itself. The internal resistance of the primary cell depends upon the size of the electrodes, the distance between them in the solution, and the resistance of the solution. The larger the electrodes and the closer together they are in solution (without touching), the lower the internal resistance of the primary cell and the more current it is capable of supplying to the load.

Note: When current flows through a cell, the zinc gradually dissolves in the solution and the acid is neutralized.

COMBINING CELLS

In many operations, battery-powered devices may require more electrical energy than one cell can provide. Various devices may require either a higher voltage or more current, and in some cases both. Under such conditions, it is necessary to combine, or interconnect, a sufficient number of cells to meet the higher requirements. Cells connected in series provide a higher voltage, while cells connected in parallel provide a higher current capacity. To provide adequate power when both voltage and current requirements are greater than the capacity of one cell, a combination series-parallel network of cells must be interconnected.

When cells are connected in **series** (see Figure 17.4), the total voltage across the battery of cells is equal to the sum of the voltage of each of the individual cells. In Figure 17.4, the four 1.5-V cells in series provide a total battery voltage of 6 V. When cells are placed in series, the positive terminal of one cell is connected to the negative terminal of the other cell. The positive electrode of the first cell and negative electrode of the last cell then serve as the power takeoff terminals of the battery. The current flowing through such a battery of series cells is the same as from one cell because the same current flows through all the series cells.

To obtain a greater current, a battery has cells connected in **parallel** as shown in Figure 17.5. In this parallel connection, all the positive electrodes are connected to one line, and all negative electrodes are connected to the other. Any point on the positive side can serve as the positive terminal of the battery and any point on the negative side can be the negative terminal.

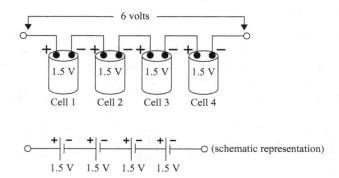

FIGURE 17.4   Cells in series.

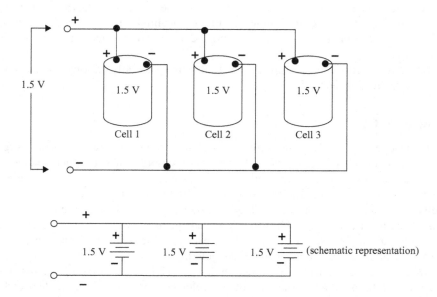

FIGURE 17.5   Cells in parallel.

The total voltage output of a battery of three parallel cells is the same as that for a single cell (Figure 17.5), but the available current is three times that of one cell; that is, the current capacity has been increased.

Identical cells in parallel all supply equal parts of the current to the load. For example, of three different parallel cells producing a load current of 210 ma, each cell contributes 70 ma.

Figure 17.6 depicts a schematic of a **series-parallel** battery network supplying power to a load requiring both a voltage and current greater than one cell can provide. To provide the required increased voltage, groups of three 1.5-V cells are connected in series. To provide the required increased amperage, four series groups are connected in parallel.

TYPES OF BATTERIES

In the past 30 years, several different types of batteries have been developed. In this text, we briefly discuss the dry cell battery and those batteries that are currently used to store electrical energy in a reversible chemical reaction as applied to renewable energy production. The renewable energy source (solar, wind, or hydro) produces the energy, and the battery stores it for times of low or no renewable energy production. The types of batteries used in this application and discussed here include the lead-acid battery, alkaline cell, nickel-cadmium, mercury cell, nickel-metal hydride, lithium ion, lithium-ion polymer batteries. Keep in mind that a battery does not create energy; instead, it stores energy. For most renewable energy applications, the preferred battery type is one that is a deep-cycle battery. A deep-cycle battery is designed to deliver a constant voltage as the battery discharges. A car starting battery, in contrast, is designed to deliver sporadic current spikes. Battery-driven vehicles, such as forklifts, golf carts, and floor sweepers commonly use deep-cycle batteries. Deep-cycle battery can be charged with a lower current than regular batteries.

- **Dry cell**—The dry cell, or carbon-zinc cell, is so called because its electrolyte is not in a liquid state (however, the electrolyte is a moist paste). The dry cell battery is one of the oldest and most widely used commercial types of dry cell. The carbon, in the form of a rod that is placed in the center of the cell, is the positive terminal. The case of the cell is made of zinc, which is the negative terminal (see Figure 17.2). Between the carbon electrode and the zinc case is the electrolyte of a moist chemical paste-like mixture. The cell is sealed to prevent the liquid in the paste from evaporating. The voltage of a cell of this type is about 1.5 V.

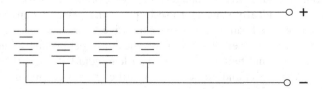

FIGURE 17.6 Series-parallel connected cells.

- **Lead-acid battery**—The *lead-acid battery* is a secondary cell—commonly termed a storage battery or rechargeable—that stores chemical energy until it is released as electrical energy.

The lead-acid battery differs from the primary-cell-type battery mainly in that it may be recharged, whereas most primary cells are not normally recharged. As the name implies, the lead-acid battery consists of a number of lead-acid cells immersed in a dilute solution of sulfuric acid. Each cell has two groups of lead plates; one set is the positive terminal and the other is the negative terminal. Active materials within the battery (lead plates and sulfuric acid electrolyte) react chemically to produce a flow of DC whenever current-consuming devices are connected to the battery terminal posts. This current is produced by chemical reaction between the active material of the plates (electrodes) and the electrolyte (sulfuric acid). This type of cell produces slightly more than 2 V. Most automobile batteries contain six cells connected in series so that the output voltage from the battery is slightly more than 12 V. Besides being rechargeable, the main advantage of the lead-acid storage battery over the dry cell battery is that the storage battery can supply current for a much longer time than the average dry cell. Lead-acid batteries can be designed to be high power and are inexpensive, safe, and reliable. Recycling infrastructure is in place for them. But low specific energy, poor cold temperature performance, and short calendar and cycle life are still impediments to their use. Advanced high-power, deep-cycle lead-acid batteries are being developed for HEV applications. However, lead-acid batteries are used for residential solar electric systems because of their low maintenance requirements and cost.

Safety note: Whenever a lead-acid storage battery is charging, the chemical action produces dangerous hydrogen gas; thus, the charging operation should only take place in a well-ventilated area.

- **Alkaline cell**—The *alkaline cell* is a secondary cell that gets its name from its alkaline electrolyte—potassium hydroxide. Another type battery, sometimes called the "alkaline battery," has a negative electrode of zinc and a positive electrode of manganese dioxide. It generates 1.5 V.
- **Nickel-cadmium cell**—The *nickel-cadmium cell*, or Ni-Cad cell, is the only dry battery that is a true storage battery with a reversible chemical reaction, allowing recharging many times. In the secondary nickel-cadmium dry cell, the electrolyte is potassium hydroxide, the negative electrode is nickel hydroxide, and the positive electrode is cadmium oxide. The operating voltage is 1.25 V. Because of its rugged characteristics (stands up well to shock, vibration, and temperature changes) and availability in a variety of shapes and sizes, it is ideally suited for use in powering portable communication equipment. Ni-Cad batteries are very expensive. Moreover, although nickel-cadmium batteries, used in many electronic consumer products, have higher specific energy and better life cycle than lead-acid batteries, they are low efficiency (65–80%) and do not deliver sufficient power and are not being considered for (HEV) applications. Cadmium is a heavy metal that is toxic

and is very expensive to dispose of, which reduces its desirability for use in hybrid application (NREL, 2009).

- **Mercury cell**—The *mercury cell* was developed as a result of space exploration activities, the development of small transceivers and miniaturized equipment where a power source of miniaturized size was needed. In addition to reduced size, the mercury cell has a good shelf life and is very rugged; they also produce a constant output voltage under different load conditions. There are two different types of mercury cells. One is a flat cell that is shaped like a button, while the other is a cylindrical cell that looks like a standard flashlight cell. The advantage of the button-type cell is that several of them can be stacked inside one container to form a battery. A cell produces 1.35 V.

- **Nickel-metal hydride**—Nickel-metal hydride batteries, used routinely in computer and medical equipment, offer reasonable specific energy and specific power capabilities. Their components are recyclable, but a recycling structure is not yet in place. Nickel-metal hydride batteries have much longer life cycle than lead-acid batteries and are safe and abuse-tolerant. These batteries have been used successfully in production electric vehicles and recently in low-volume production of HEVs. The main challenges with nickel-metal hydride batteries are their high cost, high rate of self-discharge, very high gassing/waste consumption, and heat generation at high temperatures. The need is to control losses of hydrogen, and their low cell efficiency (may be as low as 50%, typically 60%–65%) (NREL, 2009).

- **LIB or Li-ion**—The LIB are rapidly penetrating into laptop and cell-phone markets because of their high specific energy. They also have high specific power, high-energy efficiency, good high-temperature performance, and low self-discharge. Components of LIB can also be recycled. These characteristics make LIB suitable for HEV applications. However, to make them commercially viable for HEVs, further development is needed including improvement in calendar and cycle life, higher degree of cell and battery safety, abuse tolerance, and acceptable cost (NREL, 2009)—it can be said that all of these improvements are a work in progress and progress is being made on a constant basis.

- **Lithium-ion polymer batteries**—Lithium-ion polymer batteries with high specific energy (i.e., high energy per unit mass), initially developed for cell phone applications, also have the potential to provide high specific power for HEV applications. The other key characteristics of the lithium polymer are safety and good cycle life. The battery could be commercially viable if the cost is lowered and higher specific power batteries are developed (NREL, 2009).

REFERENCE

NREL (2009). *Ultracapacitors*. National Renewable Energy Laboratory. Accessed 01/29/10 @ http://www.nrel.gov/vehiclesandfuels/energystorage/ultracapacitors.html?print.

18 Lithium Batteries

INTRODUCTION

Have you ever noticed that we hardly ever think about batteries and when we do think about batteries it is when they fail. Batteries store energy but there are significant differences in how they work. Also, not only are the various types of batteries different but they also differ in effectiveness depending on different applications. The lithium-ion batteries are different than standard lead acid type batteries. One of the major differences between types of batteries is their deep discharge cycles. As mentioned earlier, lithium has a number of uses but one of the most important and valuable as component of high energy-density rechargeable lithium-ion batteries. It is concern over carbon dioxide footprint and increasing high cost for hydrocarbon fuel and its reduced supply, lithium has become more important in large batteries for powering all-electric and hybrid vehicles. It takes 1.4–3.0 kg of lithium equipment (roughly 7.5–16 kg of lithium carbonate to support a 40-mile trip in an electric vehicle (EV) before requiring recharge (Goonan, 2012). As you might imagine, this has created a large and increasing demand for lithium. Estimates of future lithium demand vary, based on numerous variables including the availability of lithium. Other variables include the potential for recycling (see Figure 18.1), widespread acceptance of EVs, or the possibility of incentives for converting to lithium-powered motors.

FOOD FOR THOUGHT

Note that increased electricity use not only will make electricity more expensive but not available in many locations and especially so in remote locations; this is the inevitable situation when more EVs are on the highways is almost a certainty and with the present state of USA's electrical grid system. Moreover, the more EV's take over the highways, the less need there will be for fossil fuels meaning that the price of gasoline and diesel will surely decrease—ironically, this will make hydrocarbon fuels more desirable.

In 2021, more than 10% of worldwide lithium reserves, expressed in terms of contained lithium, were reported to be within hard rock mineral deposits and 90%, within brine deposits. Most of the lithium recovered from brine came from Chile (1,500,000 kg, gross weight) with smaller amounts from China (more than 9,200,000 kg, gross weight), Argentina (more than 2,200,000 kg, gross weight), and the United States (750,000 kg, gross weight) (USGS, 2022). Australia and Chile also have lithium mineral reserves. Another important source of lithium is from recycled batteries. Now that the trend is to use lithium-ion batteries to power vehicles (EVs), it is expected that battery recycling will increase because vehicle battery recycling systems can be used to produce new lithium-ion batteries.

DOI: 10.1201/9781003387879-22   **193**

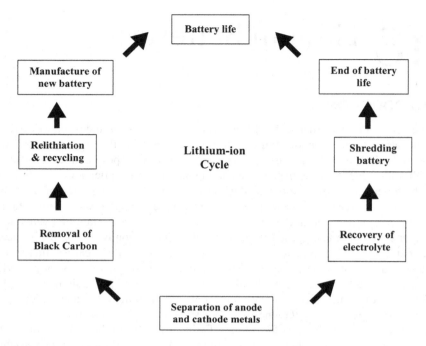

FIGURE 18.1 Lithium-ion cycle.

DID YOU KNOW?

Lithium was discovered by Johan August Arfwedson in 1817.

THE 411 ON LITHIUM-BASED BATTERIES

Recall that lithium is the lightest metal and the least dense solid and, in the latter part of the 20th century, became important as an anode material (negative pole of the battery) in lithium batteries. What makes lithium so valuable is its high electrochemical potential in high energy-density rechargeable lithium-ion battery applications. Note that the other battery metals included in lithium-ion batteries are cobalt, manganese, nickel, and phosphorus. In advanced societies, batteries are everywhere powering vehicle operations, sensors, computers, electron, and medical devices, and for electrical grid-system load-leveling and are produced and discarded by the billions each year. The concern that the demand for battery metals is increasing is real and may lead to a shortage of these metals. The metal of particular interest is lithium because it is the likelihood of the battery metals to be replaced by substitution because it has the highest charge-to-weight ratio, which is desired bore batteries in transportation applications (Goonan, 2012).

At the present time (2023), lithium batteries are enjoying a very sizeable market, powering laptop computer, cordless heavy-duty power tools, and hand-held electronic devices. Based on present and far-seeing future potential acquisitions, the graphic curve for lithium as a component of electric and hybrid vehicle batteries and for alternative energy production the graph line is almost off the charts. The concerns about the environment, about the carbon dioxide footprint of hydrocarbon-based powerplants and internal-combustion-powered vehicles, the projected hydrocarbon shortage (meaning prices are going to rise) in coming years, andthe United States dependency on foreign hydrocarbon fuels have spurred great interest in alternative energy sources. Electric-powered vehicles are beginning to take a market share from internal-combustion–powered vehicles now and in the future. The batteries required at the present time are large and will continue to be needed for power all-electric and hybrid vehicles and also for load leveling within solar- and wind-powered electric generation systems. Research on lithium for use in large batteries for electric/hybrid vehicles is on-going and making almost daily advances/improvements. Light vehicles in the future are likely to be powered by electric motors with large, light-weight batteries, and lithium is the key because of its high charge-to-weight ratio (Goonan, 2012).

The bottom line: One thing seems certain, the demand for lithium will continue to balloon with no deflation apparent. The record shows that the development of the lithium and battery industry has never lost pace. The demand for lithium-ion batteries grew by more than 30 times from 2000 to 2015 and is expected to grow by more than another 10 times by 2025 (Pillot, 2016; Okubo, 2019).

DID YOU KNOW?

Lithium is so light that it can float on water and is soft enough to be cut with a knife.

LITHIUM BATTERY TYPES

Note that lithium batteries contain metallic lithium and are not rechargeable. Small medical devices along with power watches and hand-held calculators usually have lithium batteries—called primary lithium batteries. These batteries provide more useable power per unit weight than do lithium-ion batteries (aka secondary batteries). Just because the various types of electronics uses lithium batteries it does not mean that they use the same type of lithium batteries. It is important to point out that lithium-ion batteries do not use pure lithium element but instead use lithium compounds, which are much more stable and less likely to oxidized spontaneously (Green Batteries, 2009).

Although there are several types of lithium-ion battery configurations, a closer look at the six main types of lithium batteries is provided here. Keep in mind that many of these batteries are not generally available in standard household sizes but rather are manufactured specifically for a particular electronic device (Goonan, 2012).

In operation, lithium batteries rely on lithium ions to store energy by creating an electrical potential difference (voltage) between the negative and positive poles of the battery. Inside the battery is a separator, an insulator which divides the two sides of the battery blocking the electrons while still allowing the lithium ions to pass through. When charging, lithium ions move from the positive side of the battery to the negative side through the separator. During discharge of the battery, the irons move in the reverse direction. It is the movement of lithium ions that causes the electrical potential difference (voltage). When a load (electronic device) is connected to a lithium battery, the electrons blocked by the separator are forced to pass through the device and power it.

The different types of batteries depend on unique active materials and chemical reactions to store energy—remember, batteries store energy and not electricity. The different types of lithium batteries get their names from their active materials.

Lithium Iron Phosphate Battery

The lithium iron phosphate battery (LFP—lithium ferro-phosphate) is a type of lithium-ion battery using lithium iron phosphate ($LiFePO_4$) as the cathode material (negative pole).and a graphic carbon electrode with a metallic backing as the anode (positive pole). Because of their lower cost, high safety (they do not explode with overcharging, nor produce flammable gases, period), low toxicity (not usually consider hazardous waste), long cycle, and other factors, the lithium iron phosphate cells ($LiFePo_4$) are generally accepted as the best lithium-ion battery for industrial applications. LFP battery weighs less than 33% to 25% of the weight of a lead-acid battery of equivalent power.

It was pointed out above that the batteries differ in the number of deep discharge cycles they can deliver. The deep cycle battery often looks like a standard car battery; however, they are quite different. A deep cycle battery is a battery designed to provide sustained power over a long period and run reliably until they are 80% discharged or more, at which point it needs to be recharged. In the case of LFP batteries, they can deliver more than 5,000 deep discharge cycles, compared to around 300–800 for 10-year design-life VRLA (value-regulated lead acid), or 1,500 cycles depth of discharge for 20-year design-life VRLA (IRENA, 2015). LFP battery cells have a nominal voltage of 3.2 V—connecting four of them in series as a battery yields 12.8 V. The LFP can, in higher discharge-rate applications, produce double the usable capacity of similarly rated lead-acid batteries. Moreover, in comparison to lead-acid batteries, which experience "voltage sag," the LFP batteries have a flat voltage discharge curve and have a higher discharge-rate capability. Another important difference between the standard lead-acid-type battery and the LFP is that the LFP can remain partially discharged state for extended periods without causing permanent reduction of capacity. The LFP battery types also do not suffer from thermal runaway.

LFP batteries have both pros and cons, benefits and drawbacks. On the benefit side is the ability of LFP batteries to provide a large amount of power and they are durable with a long life cycle and safety. On the drawback side is the LFPs' low specific energy (i.e., the ability to deliver a high current and indicates lading capability). Another drawback is low performance during low temperatures—making this battery not a great fit in some high cranking applications.

Lithium cobalt oxide

Lithium cobalt oxide (LCO) ($LiCoO_2$) batteries do not perform well in high-load applications because they have a high specific energy but low specific power. The good news is that they can deliver power over a long period under low-load applications. LCO batteries are commonly used in small portable electronic devices such as mobile phones, tablets, laptops, cameras, and other devices with the list of uses increasing daily. The truth be told, however, the LCO battery is losing popularity to other types of lithium batteries mainly to the cost—cobalt is not cheap; they are not cost-effective—and there are also concerns about personal safety (they have low thermal stability).

Lithium manganese oxide

A lithium manganese oxide (LMO) battery is a lithium-ion cell that uses manganese dioxide, MnO_2, as the cathode (positive pole of battery). This chemistry creates a three-dimensional structure that improves ion flow and functions through the same intercalation/deintercalation mechanism as other commercialized secondary battery technologies. The LMO batteries have low internal resistance and increase current handling and, at the same time, improve thermal stability and safety. Currently, LMO batteries are used for medical instruments, some portable power tools, and some hybrid and EVs. The primary benefit derived from using LMO batteries is that they charge quickly and offer high specific power—they are very flexible and they provide higher current than LCO batteries and also operate safely at higher temperatures. The main downside of LMO batteries is their short lifespan (only 300–700 charge cycles) which is significantly less than other lithium battery types.

Lithium nickel manganese cobalt oxide

Lithium nickel manganese cobalt oxide (NI-NMC, LNMC, or NMC) batteries combine the benefits of the three main elements in the cathode: nickel (high specific energy but is not stable alone), manganese (exceptionally stable but low in specific energy), and cobalt. When these elements are combined, they yield stable chemistry with a high specific energy. NMC batteries are similar to LMO batteries in that they both used in power tools as well as electronic powertrains for e-bike, scooters, and some EVs. NMC batteries have higher thermal stability than LCO batteries; this characteristic makes them safer overall. Moreover, because NMC batteries have high energy density and a longer life cycle at a low cost they are advantageous for use as compared to other cobalt-based batteries. On the other hand, one significant drawback of NMC batteries is that they have a slightly lower voltage than cobalt-based batteries.

Lithium nickel cobalt aluminum oxide

Lithium nickel cobalt aluminum oxide (Li-NCA, LNCA, NCA) batteries offer high specific energy with better than average specific power and long life cycle. All this adds up to the ability to deliver relatively high amounts of current flow for long periods. Because of these advantages, that is, the ability to perform high-load applications with longer battery life makes NCA batteries the battery of choice in the EV market (it is the battery of choice for Tesla). The main drawbacks of the NCA batteries is their high cost and they have an issue with being less safe than other lithium-ion batteries.

Lithium titanate

To this point, the preceding types of lithium-ion batteries where the cathode has been presented, however this is not the case with the lithium titanate (LTO) battery. In the LTO battery, the graphite in the anode has been replaced with LTO and use LMO or NMC as the cathode chemistry. Experimentation and experience demonstrate that the LTO battery is faster to charge than the other lithium-ion batteries. Moreover, LTO batteries are safe and have a long lifespan and an extremely wide operating temperature. The benefit of using LTO batteries in EVs and charging station is uninterrupted power supplies, and military and aerospace equipment is well documented—and appreciated. Okay, on the other side of the good news always lies some bad news and with the LTO battery all the news is not exactly positive. As a case in point, consider that the LTO battery is a low-energy density-type battery—it stores a lower amount of energy relative to its weight when compared to some other lithium batteries. The icing on the cake, so to speak, is that the LTO battery is very expensive.

DID YOU KNOW?

While it is true that lithium-ion batteries are quickly gaining in popularity, it is also true that the standard lead-acid battery for vehicles is still the flagship type used by the majority of vehicle manufacturers at the present time. The point is lithium-ion batteries are the batteries of the future and the standard lead-acid type of battery is going on the way of the dinosaur and gooney birds.

The types of lithium-ion batteries discussed above are (at the present time) not generally available in standard household sizes but instead are manufactured for a particular electronic device. It does seem inevitable that lithium-ion batteries of one type or another will surge ahead of standard lead-acid storage batteries in common usage and become the basis for the future EV power supply. Note that the major difference between batteries for electronics and batteries for EVs will be and is size. Increased size is obtainable by making assemblies of small cells or by developing singular large cells (Goonan, 2012).

LITHIUM CELL TYPES

We have covered the different types of lithium-ion batteries now it is time to shift gears, so to speak, and to describe the three types of lithium cells: cylindrical, prismatic and pouch cells—note that the cylindrical and prismatic cells are the popular cells.

CYLINDRICAL CELL

Cylindrical cells are very popular because they are easy to manufacture and mechanically stable. This cell's advantages also include in its thin profile, which is an effective use of space. Cylindrical cells achieve increased flexibility due to their rectangular shape that facilitates better layering. This type of cell is generally used in portable technologies such as in wireless communications, laptops, and medical devices. Note that the larger cylindrical cells are the cells of choice for EVs; Tesla is catalyst for increasing market demand for this type. But it is important to note that increasing market demand for the cylindrical cell is also affecting the demand for cells found in mobile phones, tablets, and other lightweight electronic devices. They're also safe; that is, if the internal pressure grows too great, most cells are deliberately designed to rupture which mitigates safety hazards. Also, these cells offer longevity and they are economical (Buchmann, 2004).

PRISMATIC CELL

The prismatic cells shares many of the advantages of the cylinder cells. These cells have a relatively long history due to their development in the early 1990s. These cells are made in various sizes and capacities and custom made for cell phones (Buchmann, 2004). Their negatives include a sensitive to deformation in high-pressure situations and they are available in limited standardized sizes.

POUCH CELL

The pouch cell is a relatively new (1995) radical design that allows tailoring the cell to the exact dimensions of the electronic device manufacturer. Their best feature is their ability to be easily assembled into battery packs as needed (Buchmann, 2004), giving them a packaging efficiency of roughly 95%. From a safety point of view when the reduced weight of a non-metal casing not only reduce the chance for explosive (package swells instead) but from a practical perspective the weight is reduced. These cells are used in the military, automotive, and other consumer applications.

REFERENCES

Buchmann, I. (2004). *Battery packaging—A look at old and new systems*. Battery University. Accessed 12/17/22 @ http://www.batteryuniversity.com/print-partone-9.htm.
Goonan, T.G. (2012). Lithium use in batteries. U.S. Geological Survey Circular 1371, 14 p. Available @ http://pubs.usgs.gov/circ/1371/.
Green Batteries (2009). *Lithium-ion battery frequently asked questions*. Green Batteries. Accessed 12/19/22 @ http://www.Greenbatteries.com/libafa.html.
IRENA (2015). Renewables and electricity storage. Accessed 12/14/22 @ https://www.irena.org/publications/2015/uun/renewables-and-electricity-stoarage.
Okubo, M. (2019). Creating a future energy world on the foundation of technology and innovation. Accessed 12/12/22 @ https://www.japantimes.co.jp/country-report/2019/06/28/north-rhine-westphalia-report;2019.
Pillot, C. (2016). The rechargeable battery market and main trends 2015–2025. Accessed 12/12/22 @ https://www.slideshare.net/christorphPILLOT/c.pillot-june-2016-64406919.
USGS (2022). Lithium. Accessed 12/13/22 @ https://pubs.usgs.gov/periodicals/mcs2022-lithium.pdf.

19 Lithium-Ion (Rechargeable) Battery Production

SUPPLY CHAIN

Until recently, specifically prior to Covid-19 and the invasion of Ukraine, most people had paid little attention to any type of supply chain and even fewer could define or had a clue what the supply chain for anything is all about. Moreover, along with disease and geopolitical uncertainty (e.g., invasion of Ukraine), other issues can and do affect supply chains, of any type. For example, increasing cost of living as a result of skyrocketing inflation along with energy shortages, labor unrest, and extreme weather all can and do affect supply chains, of any type.

What is about the supply chain for lithium material for the production of lithium-ion batteries?

Certainly, the issues with supply chain malfunction including finding a reliable supply, slow delivery, or no delivery at all can affect the lithium supply chain. In order to understand (or foresee) any problems with lithium's supply chain, it is first important to visually look at an outline of the current lithium-battery supply chain (see Figure 19.1) from upstream raw materials production to midstream processing and finally to downstream end-of-recycling.

Referring to Figure 19.1, the upstream section (raw materials production) of the lithium-based battery supply chain includes mining and extraction of materials needed for manufacturing lithium-ion batteries. These materials include lithium, nickel, cobalt, and graphite. The midstream section (processing and cell manufacturing) in the lithium-ion battery supply chain includes materials processing and cell manufacturing including processing materials to battery-grade, cathode and anode powder production, separation and electrolyte production, and electrode and cell manufacturing. The downstream section includes pack manufacturing and end-of-life recycling and reuse. The manufactured end product, lithium-ion batteries, are

---

DID YOU KNOW?

Note that increasing preferences, decreasing battery costs, environmental regulation, increasing fuel costs, new end-use markets, and technological advances have all play a role in the changing consumption and substitution patterns of rechargeable batteries, particularly in automotive and consumer electronic product applications (Goonan, 2012).

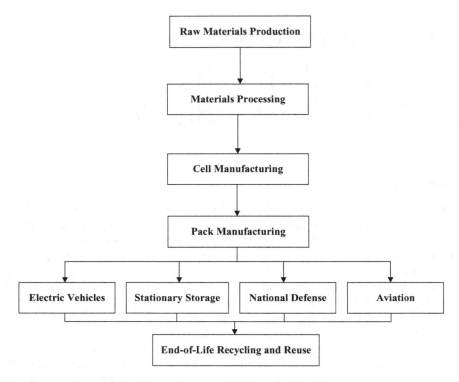

FIGURE 19.1 The current lithium-battery supply chain.

used in aviation, national defense, stationary storage, and electric vehicles. After use, lithium-ion batteries are recycled for reuse or they are designated as waste products and handled accordingly.

BATTERY PRODUCTION

Lithium batteries were first entered the market in 1993. Before 2012, lithium-ion battery (rechargeable) production in the United States was limited to small-scale, high-profit-margin niche markets, such as space applications, medical, or military applications. In 2008, the greater part of general-use lithium-ion batteries were produced in Japan (the major producer of lithium-based batteries with the majority of the batteries lithium-ion batteries), the Republic of Korea, and China (Wilburn, 2008). General Motors, in 2009, announced the construction of a lithium-ion battery pack production to be located in Warren, Michigan and at the time of this writing, General Motors lithium-ion battery plant is in the construction phase with focus on manufacturing batteries of electric vehicles. The lithium battery market between 1993 and 2008 consisted of primary (non-rechargeable) and secondary (rechargeable) batteries for electronic devices. Note that in 2008, only about 0.2% of lithium produced went to automobile batteries (Wilburn, 2008). With the current movement toward replacing hydrocarbon fuels in transportation vehicles, the percentage of lithium produced currently (2022) is on the increase with usage steadily increasing.

BATTERY RECYCLING

When lithium-ion vehicle batteries reach their end-of-life, they need to be (and are being) recycled. In 2009, an estimated 3,700 tons of lithium, contained scrap batteries, became available to the global market. The truth be told is that there are hidden dangers tied to improper handling and disposal of batteries at their end-of-life. The problem is that many people are unaware of the dangers, which are causing an increasing fires at recycling centers, waste facilities, and garbage trucks, all of which result in millions of dollars in physical damage and putting lives in danger.

All batteries can and should be recycled. Certain types of batteries, such as nickel cadmium rechargeable, can contaminate the environment if not properly disposed. The point is that these batteries contain valuable components that can be recycled—parts of batteries can be reclaimed and put into other products.

The bottom line: In regards to safety, we can play a role in reducing the potential risks to people, property, and the environment. Isn't this what recycling is all about?

REFERENCES

Goonan, T.G. (2012). Lithium use in batteries: U.S. Geological Survey Circular 1371, 14 p. Available @ http://pubs.usgs.gov/circ/1371/.

Wilburn, D.R. (2008). Material use in the United States—Selected case studies for cadmium, cobalt, lithium, and nickel in rechargeable batteries. U.S. Geological Survey Scientific Investigations Report 2008-5141, 18 p. Accessed 12/20/22 @ http://pubs.usgs.gov/sir/2008/5141.

20 Lithium-Ion Battery Operation

HOW DO THEY WORK?

To this point in the book, it has been shown that lithium-ion batteries are very popular and their popularity continues to grow with applications in laptop, PDAs, cell phones, and iPods. They're so common because, pound for pound, they're the most energetic rechargeable batteries available at the present time. So the question is what makes lithium-ion batteries so energetic and so popular?

Lithium-ion batteries have important advantages over competing technologies and with the present move underway to switch from hydrocarbon fuels for transportation to electric vehicles, the importance of lithium-ion batteries is steadily increasing—all of this is based on the advantageous capabilities of the batteries. For example, they are much lighter than other types of rechargeable batteries of the same size. In the manufacture of electric vehicles where the battery is basically the undercarriage of the vehicle, the lighter weight advantage of the lithium-ion-type batteries is important. In addition to their being lighter than batteries of the same size, the electrodes of lithium-ion batteries are made of lightweight lithium and carbon. Another advantage of the lithium-ion battery is its ability to store a significant amount of energy in atomic bonds, which means that the lithium-ion battery is a very high energy density battery. Another significant advantage of the lithium-ion battery is their ability to hold its charge. Not only can lithium-ion batteries hold a charge they also have no memory effect meaning that there is no need to fully discharge them before recharging. Finally, lithium-ion batteries can handle numerous charge/discharge cycles.

LITHIUM-ION BATTERY CHEMISTRY

All lithium-ion batteries look pretty the same on the inside but the battery packs come in various sizes and shapes. A good example to look at internally is the laptop battery pack. Recall that lithium-ion cells can be either cylindrical and look a lot like AA cells, or they can be prismatic (either square or rectangular). Inside the laptop itself is one or more temperature sensors that monitor battery temperature. To maintain safe levels of voltage and current, a voltage converter and regulator circuit are included. Power and information in and out of the battery pack are provided by shielded notebook connector—shielding the cable without shielding the connector is only doing half the job. Monitoring the energy capacity of individual cells in the battery pack is accomplished using a voltage tap. To ensure that the battery is charged as quickly and fully as possible, a battery charge state monitor (a small, sophisticated computer) is included. This battery stage charge monitor is important and draws its power from

DOI: 10.1201/9781003387879-24

the batteries. Because it does draw power from the batteries, the lithium-ion batteries lose a small percent of their power when sitting idle.

Note that the computer will shut down the flow of power anytime the battery pack gets too hot during use or charging. Too much heat is the computer killer, so to speak. For example, if you work with a computer outdoors in the Sun in hot weather or leave the computer sitting anywhere where there is excessive heat, the computer will not operate until it cools off and it might not operate at all or power up until it cools off. The cells are ruined, the computer is shut down, whenever the cells become totally discharged. The battery charge state monitor may also keep track of the number of charge/discharge cycles that send out information so the laptop's battery meter can inform about how much charge is left in the battery.

BATTERY CELL MAKEUP AND OPERATION

In order to illustrate a simple rendition of the lithium-ion battery in operation, Figure 20.1 is provided. In this simplified example, you can see that the battery is made up of an anode, cathode, and electrolyte. In practice, the cell also includes separators and two current collectors (positive and negative). The lithium is stored in the anode and cathode. The electrolyte carries positively changed lithium ions from the anode to the cathode and vice versa thought the separator. It is the lithium ions that create free electrons in the anode which creates a charge at the positive current collector. It's all about creating current flow and this current flows from the current collector though a device, the load (cell phone, computer, etc.), to the negative current collector. The separator blocks the flow of electrons inside the battery.

DID YOU KNOW?

Energy density [measured in watt-hours per kilogram (Wh/kg)] and power density [measured in watts per kilogram (W/kg)] are the two most common concepts associated with batteries.

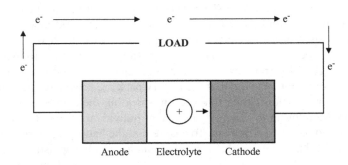

FIGURE 20.1 Simplified drawing of lithium-ion operation.

While the battery discharges and provides an electric current (electron flow), the anode releases lithium ions to the cathode, generating a flow of electrons (current flow) from one side to the other. Note that when plugging in the device, the opposite happens: Lithium ions are released by the cathode and received by the anode.

The bottom line: Knowing how a lithium-ion battery works/operates is important for those who are interested in this type of battery mainly because these batteries have several applications and findings of their potential use increases almost daily. Although the focus in this presentation has been wide ranging and aimed at lithium-ion applications in all areas, it is the use of the lithium-ion batteries in electric vehicles that is the main focus of this book. Thus, in the next chapter, foundational material is provided about powering vehicles by electricity instead of by fossil fuels.

Part 5

Environmental and Safety Concerns

21 Electric Vehicle Terminology

ELECTRIC VEHICLE KEY TERMS

When a lithium-ion or lithium-ion polymer battery is used to power a BEV (battery-powered electric vehicle) or HEV (hybrid electric vehicle), in many cases, the terminology used or applied is the same. The key terms currently used are listed and defined in the following. Other terms not defined in the following are defined when introduced.

- **Alternating current (AC)**—a charge of electricity that periodically changes directions.
- **All-electric range (AER)**—the range an electric vehicle (EV) is able to reach safely using electricity.
- **All-electric vehicle (AEV)**—a BEV.
- **Combined Charging System (CCS)**—a standard, fast charging system (50 kw and up), often used with the SAE J1772 (J-Plug).
- **CHAdeMO**—is the trade mark of a standard, fast charging system (50 kW and up).
- **Connector**—a device attached to a cable from an Electric Vehicle Supply Equipment (EVSE) that converts to an EV allowing it to charge.
- **Direct current (DC)**—a charge of electricity that flows in one direction and is the type of power that comes from a battery.
- **DC fast charging**—is the fastest (high-powered) way to charge EVs—can charge a battery to 80% in 30 min and then slows in order to not overheat the battery.
- **Energy density vs power density**—energy density is measured in watt-hours per kilogram (Wh/kg) and is the amount of energy the battery can store with respect to its mass. Power density is measured in watts per kilogram (w/kg) and is the amount of power that can be generated by the battery with respect to its mass.
- **Extended-range electric vehicles (EREVs)**—have the ability to run on a small internal combustion engines (called range extender) if the battery gets low. Usually, the range extender powers a generator that transfers electricity to the batteries and motor.
- **Fuel cell electric vehicle (FCEV)**—uses compressed hydrogen gas for fuel.
- **Full hybrid electric vehicle (FGFV)**—combines a conventional internal combustion engine system with an electric propulsion system.
- **Level 1 (slow) charging (L1)**—every EV comes with a universally compatible L1 charge cable that plugs into any standard grounded 120-V outlet.

DOI: 10.1201/9781003387879-26

The L1 charger power rating tops out a 2.4 kW, restoring about 5–8 mph charge time, about 40 miles every 8 hours. Many drivers refer to the L1 charge cable as a trickle charger or emergency charger. The L1 charger will not keep up with long commutes or long drives anywhere.

- **Level 2 (fast) charging (L2)**—this charger runs at higher input voltage, 240 V and is usually a dedicated 240-V circuit in a driveway or garage that can be used with a J-plug connector. This is the most common charging system for residential use and can be found at commercial facilities. These chargers tend to top out at 12 kW, restoring up to 12–25 mph charge, about 100 miles every 8 hours.
- **Level 3(rapid) charging (L3)**—is the fastest EV charger available; it charges a battery to 80% in 30 minutes (using 480-V circuits) and then slows to prevent overheating of the battery. Both CHAdeMO and SAE CCS connectors are used.
- **Molten salt battery**—uses molten salts as an electrolyte.
- **MPGe—million per gallon equivalent**—used to compare the fuel efficient of EVs and internal combustion engines. It is determined by measuring how far an EV can travel on 33.7 kWh (the energy equivalent of one gallon of gas).
- **Range**—distance an EV can travel on pure electric power before the battery requires a recharge. **Range anxiety or range worry** occurs whenever concern arises that the EV's battery power will run out before the destination is reached.
- **Regenerative braking**—used in EVs to transfer energy from the braking function to the battery for stored energy.
- **SAE J1772 (J-Plug)**—standard North American electrical connection for EVs—works with Level 1 and Level 2 systems.

ONE MORE WORD

This chapter has presented information on EVs and specifically on battery basics and batteries presently used in EVs. In the next chapter, we add another building block, vehicle dynamics involved in the motion of vehicles, as foundation that is involved with the electric powered vehicle. The information provided is based on F. Spellman's (2023) text.

REFERENCE

Spellman, F. (2023). *The science of electric vehicles: Concepts and applications*. Boca Raton, FL, CRC Press.

22 Dynamics of Vehicle Motion

INTRODUCTION[1]

In the engineering and design of vehicles-on-wheels, vehicle dynamics (refers to an electric vehicle, EV) is the study of the vehicle in motion and how it behaves in motion. It is important for design and engineering professionals to fully understand what a vehicle does and basically what does it add up to. Essentially, EVs work in generally the same way ones powered by gas, diesel, biodiesel, and hydrogen. There is a fuel source, a drive unit, and a gearbox to provide forward and backward motion. In short, an EV vehicle-on-wheels includes subsystems/modules as follows.

- **EV power module**—includes EV traction, Electronic Control Unit (ECU), Transmission Control Unit (TCU), motor, gear box—single-speed transmission, drive axles.
- **EV chassis module**—includes suspension, steering, braking and parking, tires, and wheels.
- **EV body module**—includes bonnet, doors, roof, trims, and so forth.

Let's get back to vehicle dynamics (aka vehicle mechanics—vehicle dynamics is a subset of engineering dealing with and based on established mechanics). Simply, vehicle dynamics for vehicles-on-wheels is the study of vehicle motion. More specifically, vehicle dynamics is the study of how a vehicle's forward movement changes in response to driver inputs, propulsion system outputs, ambient conditions, air/surface/ water conditions, and so forth. Simply, and to the point, vehicle dynamics is all about the study of how much energy is required to propel vehicles-on-wheels and the limiting factors involved in determining the amount of energy required to move the vehicle.

Again, it is all about motion. Okay, that is logical but what are the factors affecting vehicle dynamics?

The factors affecting vehicle dynamics are many and varied, including drivetrain and braking, suspension and steering, distribution of mass, aerodynamics, and tires. In the vehicles-on-wheels application, the fundamentals of vehicle design are paramount and embedded in basic physics (mechanics of). With regard to pure science impact, vehicle dynamics is all about Newton's second law of motion that states *the acceleration of an object is proportional to the net force exerted on it.* Okay, in this presentation, the term "net force" refers to the amount of force acting on the vehicle-on-wheels.

DOI: 10.1201/9781003387879-27

FACTORS AFFECTING VEHICLE DYNAMICS

As mentioned, the factors affecting vehicle dynamics are many and varied, including drivetrain and braking, suspension and steering, distribution of mass, aerodynamics, and tires. The dynamics that govern vehicle motion based on the forces acting on a rolling vehicle including aerodynamic drag, rolling resistance, hill climbing, linear and angular acceleration, and tractive force. Let's look at each of the factors affecting vehicle dynamics.

VEHICLE-ON-WHEELS LAYOUT

In the drivetrain and braking factor category of vehicle dynamics vehicle-on-wheels layout is derived from the location of the engine and the drive wheels. The layouts can be divided into the three categories of front-wheel drive (FWD), read-wheel drive (RWD), and four-wheel drive (4WD). In practice, the many different combinations of engine location and driven wheels actually employed are dependent on the application for which the vehicle-on-wheels will be used.

POWERTRAIN

Simply, the powertrain consists of the units that provide power to the wheels of the vehicle.

BRAKING SYSTEM

The braking system is composed of mechanical devices that restrain motion by absorbing energy from the moving system.

GEOMETRY OF SUSPENSION, STEERING AND TIRES

- One of the considerations in suspension and steering systems in vehicles-on-wheels is steering geometry, namely the *Ackermann steering geometry*; it is a consideration which is focused on the geometric arrangement of linkages used to steer the vehicle-on-wheels and the intention is to avoid the need for tires to slip sideways when following whatever the path around a curve. The geometric solution is for all wheels to have their axles arranged as radii of circle with a common center point (Norris, 1906). Okay, what does all this mean? It means that in "turntable steering" as the rear wheels are fixed, the center point must be on line extended from the rear axle. Note that intersecting the axes of the front wheels on this line as well requires that the inside front wheel be turned, when steering, through a greater angle than the outside wheel (Norris, 1906).

  Instead of the preceding turntable steering, where both front wheels turned around a common pivot, each wheel gained its own pivot, close to its own hub. While more complex, this arrangement enhances controllability by avoiding large inputs from road surface variations being applied to the end of a long lever arm, as well as greatly reducing the fore-and-aft travel of the steered wheels (Norris, 1906).

Note that modern vehicles-on-wheels do not use Ackerman steering, even though the principle is good for slow-speed maneuvers but it ignores important dynamic and complaint effects.

- *Axle track* in vehicles-on-wheels refers to those vehicles having two wheels on an axle; it is the distance between the hub flanges on an axle (Car Handling Basics, 2022). Track refers to the distance between the centerline of two wheels on the same axle. Axle and track are commonly measured in millimeters or inches (BMW M3 E46, 2022).
- *Camber angle* is one of the angles made by the wheels of a vehicle; stated differently, it is the angel between the vertical axis of a wheel and the vertical axis of the vehicle when viewed from the front or rear.
- *Caster angle* causes a wheel to align with the direction of travel. Caster displacement moves the steering axis ahead of the axis of rotation.
- *Ride height* or *ground clearance* is the distance or space between the base of the vehicle tire and the lowest point of the vehicle (usually the axle).
- *Roll center* of a vehicle is the notional point at which the concerning forces in the suspension are reacted to the vehicle body.
- *Scrub radius* is the distance at the road surface between the tire center line and the steering axis inclination.
- *Steering ratio* is the ratio between the turn of the steering wheel (in degrees) or handlebars and the turn of the wheels (in degrees). For electric motorcycles and bicycles, the steering ratio is 1:1, because the steering wheel is attached to the front wheel. In most electric passenger cars, the ratio is 12:1 and 20:1 (ratios 13–14 are considered fast and ratios above 18 are considered slow).
- *Toe*, in vehicles-on-wheels toe (aka *tracking*), as a function of static geometry, and kinematic and compliant effects is the symmetric angle that each wheel makes with the longitudinal axis of the vehicle
- *Wheel alignment* is sometimes referred to as breaking or tracking that consists of adjusting angles of wheels to manufacturer specifications.
- *Wheelbase* is the distance between the front and rear axles of a vehicle-on-wheels.

Some aspects of vehicle dynamics are due to mass and its distribution. Mass distribution is the spatial distribution of mass within a solid body. These include the following:

- *Center of mass* is the unique point where the weight relative portion of the distribution sums to zero.
- *Moment of inertia* depends on the moment of the very different moment of inertia depending on the location and orientation of the axis or rotation.
- *Roll moment* is a product of force and distance and causes a vehicle to roll, rotating about it longitudinal axis.
- *Sprung mass*, in a vehicle-on-wheels with a suspension, is the portion of the vehicle's total mass that is supported by the suspension, including in most applications approximately half of the suspension itself.

- *Unsprung mass* sometimes called unsprung weight of a vehicle is the mass of the suspension directly connected.
- *Weight distribution* is the apportioning of weight with a vehicle-on wheels.

Some aspects of vehicle dynamics are due to aerodynamics aspects, which include the following:

- *Automobile drag coefficient* is a common measure in automotive design. Drag is the force that acts parallel to and in same direction as the air flow.
- *Automotive aerodynamics* is the study involved with reducing drag and wind noise and preventing undesirable lift in vehicles-on-wheels.
- *Center of pressure* is the point where the total sum of a pressure field acts on a body causing a force to act through that point.
- *Downforce* is a downward lift force created by the aerodynamic features of a vehicle-on-wheels.
- *Ground effect (automobiles)* is a series of effects which have been exploited in automotive aerodynamics to create downforce.

Vehicle dynamics is directly affected by the tires of a vehicle-on-wheels. For instance, one of the interesting factors affecting vehicle dynamics is known as the *Magic Formula* tire models. Developed by Hans Pacejka, the Magic Formula actually consists of a series of formulae that Pacejka developed over the last 20 years.

The significance of the Magic Formula?

Well, the truth be told, the Magic Formula is widely used in professional vehicle dynamics simulations because they are reasonably accurate and very easy to program and more importantly they solve quickly.

Used for what?

Good question. The Magic Formula is, as previously stated, a series of tire design models that Pacejka developed over time.

So, what is so magical about the Magic Formula?

First off, there is no particular physical basis for the structure of the equations chosen, but they fit a wide variety of tire constructions and operating conditions—in short, the Magic Formula is not only easy to use but is adaptable to several different applications or requirements and is widely used in professional vehicle dynamics simulations and they are reasonably accurate and easy to program and solve quickly (Plasterk, 1989).

Along with the Magic Formula, there are other aspects of vehicle dynamics related to tires include camber thrust, circle of forces (i.e., a useful way of thinking about the dynamic interactions between the vehicle's tire and road), contact patch (i.e., the pneumatic touch of the tire to road surface), cornering force, ground pressure, pneumatic trail (i.e., the trail of the tire), and radial force variation (i.e., road force variation is the property of a tire that affects steering, traction, braking, and load support) (Cortez, 2014). A property of pneumatic tires that describes the delay between when a slip angle is introduced and when cornering force reaches its steady-state value is known as *relaxation length* (Pacejka, 2005). *Rolling resistance* is the force resisting the motion when a body (tire) rolls on a surface. The torque a tire develops as it rolls along is known as *self-aligning torque* (*aka aligning torque, aligning moment, SAT, or MS*) it tends to steer it, that is, rotate it around is vertical axis. *Skid* occurs when one or two tires slip relative to the road. *Slip angle or sideslip* is the angle between the direction in which a wheel is pointing and the direction in which it is actually traveling (Pacejka, 2005). Another characteristic in vehicle dynamics related to tires is *slip* which is the relative motion between the tire and the road surface it is moving on. A subset of slip is *spinout* which occurs when a vehicle rotates in one direction during a skid. *Steering ratio* refers to the ratio between the turn of the steering wheel (in degrees) and the turn of the wheel (in degrees) (Pacejka, 2005). The behavior of tires under load is called *tire load sensitivity*.

PURE DYNAMICS

Beyond distribution of mass, aerodynamics, and tires, some attributes and aspects of vehicle dynamics are purely dynamic. For example, *body flex* is a lack of rigidity in a vehicle-on-wheels chassis. The axial rotation of a vehicle-on-wheels toward the outside of a turn called *body roll* is another example of a purely dynamic aspect of vehicle dynamics. Another term used in vehicle dynamics for a pure dynamic attribute or aspect is *bump steer*. Bump steer is caused when one wheel falls down into a rut or hole or hits a bump causing the vehicle-on-wheels to turn itself.

Note that there are several other factors, attributes, and aspects of pure dynamics that impact vehicle dynamics and more importantly vehicle operation. A few of these other factors, attributes, and aspects of aerodynamics affecting operation of vehicles-on-wheels include directional stability, critical speed, pitch, yaw, roll, speed wobble, understeer, oversteer, weight transfer, and yaw—these are the factors that affect vehicle operation.

However, if we have to sum up the factors, attributes, and aspects of vehicle dynamics related to environmental impact on vehicles-on-wheels, it comes down to the forces acting on a rolling vehicle-on-wheels. And these forces are all about aerodynamics; at least, aerodynamics is where we begin—and we do this by describing aerodynamic drag, rolling resistance, linear acceleration, hill climbing, angular acceleration, and the total of them all, and this equates to the total tractive force required to propel the vehicle.

AERODYNAMICS[2]

It is interesting that whenever the term "aerodynamics" is mentioned anywhere at any time, it is likely that those exposed to the term are apt to equate the term with aircraft. And this makes sense when aerodynamics is thought of and defined in its simplest terms; that is, it is the way air moves around things and aircraft certainly depends on the rules of aerodynamics to enable them to fly. However, it is important to point out that anything that moves through air reacts to aerodynamics. A kite in the sky reacts to aerodynamics and aerodynamics even act on vehicles-on-wheels because air flows around cars and trucks and buses and other moving objects.

To illustrate the importance of aerodynamics and to provide foundational information for the impact of electric vehicles-on-wheels, we begin by addressing the four forces of flight—lift, weight, thrust, and drag. These forces make an object move up and down, and faster or slower. How much of each force is present changes how the object moves through the air. Let's take a closer look at each of these four forces: weight, lift, thrust, and drag.

WEIGHT

Everything around us on earth has *weight*. This force is the result of gravity pulling down on objects. To fly, an aircraft needs something to push it in the opposite direction from gravity. The weight of an object controls how strong the push has to be. Flying a kite is a lot easier than pushing upward on a jumbo aircraft.

LIFT

With regard to the push needed to move aircraft and kites upward, the push needed is called *lift*. Simply, lift is the force that is the opposite of weight. Again, in regard to aircraft, everything that flies must have lift. Again, for an aircraft to move upward, it must have more lift than weight. It is all about light air, well, with hot air balloons they have lift because the hot air inside is lighter than the air around it. The hot air rises and takes balloons with it. Helicopters lift comes from the rotor blades at the top to the helicopter. The helicopter is raised in the air via its motion in the air. However, lift for an aircraft is all about its wings.

So, the question is: How does an aircraft's wings provide lift?

The answer is: It's all about the shape of the aircraft's wings that provides the lift. Airplanes' wings are curved on top and flat on the bottom. The shape makes the air travel over the top faster than under the bottom. Consequently, less air pressure is on top of the wing. This condition makes the wing, and the airplane it's attached to, move up. Using curves to change air pressure is a trick used on many aircraft. Helicopter blades, kites, sailboats use this trick.

DRAG

Before we get into drag and its effect on electric vehicles-on-wheels, we need to get down a few definitions pertinent to the drag effect. First, we need to define drag coefficient (Cd) that is a dimensionless quantity that is used to quantify the drag or resistance of an object in a fluid environment, such as air—air is indeed classified as a fluid. Second, when addressed as vehicles-on-wheels, even though what is discussed and detailed could be applied to trains on rails, it is the vehicles that use the highways, streets, race tracks, or farm fields is what really is being addressed.

Okay, now for drag force, this is best addressed and explained via equation 22.1.

$$Fd = \frac{1}{2} pu2ACd \qquad (22.1)$$

where
Fd = drag force, which is a definition in the direction of flow
p = mass density of the fluid

u = flow velocity relative to the object
A = the reference area
Cd = drag coefficient (dimensionless)

Note: If the fluid is a gas (air), Cd depends on both the Reynolds Number (Re) and the Mach Number. Basically, the *Reynolds Number (Re)* helps predict flow pattern in different fluid flow (air flow) situations. Low Re equals laminar flow (sheet-like flow), while at high Re, flows tend to be turbulent. Mach Number (M or Ma) is a dimensional quantity in fluid dynamics—the ratio of flow velocity past a boundary to the local speed of sound.

Note that the shape of object effects on the amount of drag as shown in Figure 22.1.

In simple terms, drag can be defined as a force that tries to slow something down. Basically, it makes it hard for an object to move. Keep in mind that it is harder to run through water than through air. That is because water causes more drag than air. The shape of an object also changes the amount of drag as shown in Figure 22.1. Most round surfaces have less drag than flat ones. Narrow surfaces usually have less drag than wide ones. Simply, the more air that strikes a surface, the more drag it makes.

THRUST

The force that is the opposite of drag is *thrust*. Thrust is the push that moves electric vehicles-on-wheels forward.

ROLLING RESISTANCE

Rolling resistance (aka rolling friction or rolling drag) is the force that resists the motion of a body rolling on a surface (see Figure 22.2). The rolling resistance can be expressed in various ways but for our purposes the generic equation is shown here.

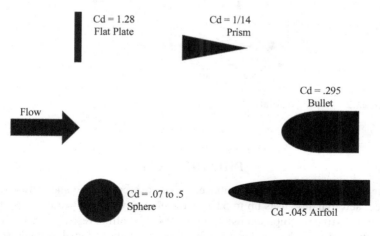

FIGURE 22.1   Shape effects on drag—all objects have the same frontal area.

Source: NASA accessed 05/31/22 @ www.grc.nasa.gov/www/Ke12/airplane/shapter.html.

$$F_r = cW \tag{22.2}$$

Where

F_r = rolling resistance or rolling fiction or rolling drag (N, lb$_f$)

c = rolling resistance coefficient—dimensionless—rolling resistance coefficient, c, is influenced by different variables like wheel design, rolling surface, wheel dimensions, and more (see Table 22.1).

$W = m\, a_g$

m = mass of the body (kg, lb)

a_g = acceleration of gravity (9.81 m/s^2, 32.174 ft/s^2)

TABLE 22.1

Some Typical Rolling Friction Coefficients

Rolling Resistance Coefficient, c	Typical Vehicle Tire Surface Conditions
0.006–0.01	Truck tire on asphalt
0.01–0.015	Ordinary care tires on concrete, asphalt, small cobbles
0.02	Car tires on tar or asphalt
0.02	Car tires on rolled gravel
0.03	Car tires on large, worn cobble
0.04–0.08	Car tire on solid sand, loose gravel, hard sand
0.2–0.4	Car tire on loose sand

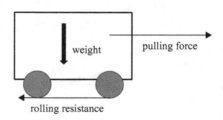

FIGURE 22.2 Rolling resistance.

DID YOU KNOW?

Note that it takes most of the usable energy to overcome drag and rolling resistance: aerodynamic uses up to 53% of the energy while drivetrain uses about 6% and auxiliary equipment uses roughly 9% and rolling resistance uses 30+% of usable energy—the good news is these losses can be reduced using presently available technology.

HILL CLIMBING

A vehicle-on-wheels traveling on a flat service at constant velocity experiences two major forces opposing the vehicle, aerodynamic drag and rolling resistance. However, if that same vehicle-on-wheels is traveling up a hill, we also need to account for gravity. One of the commonly used equations to calculate the force needed to travel up a hill is equation 22.3.

$$F_{uh} = mg \sin \psi \qquad (22.3)$$

where

F_{uh} = force required to travel uphill

mg = m is the mass of the vehicle-on-wells (in kg), g is gravity (8.91 m/s²)

ψ = the vertical angle of the road relative to flat (in radians)

Figure 22.3 illustrates why sine is the appropriate function here.

LINEAR ACCELERATION

Linear acceleration is a term related to an object in movement. Acceleration is the measure of how quickly the velocity of any moving object changes. Therefore, the acceleration is the change in the velocity, divided by the time. Acceleration has both magnitude and direction like a vector where both velocity and force are vector quantities.

Okay, acceleration and velocity have been, to a point, explained now it is time to define linear acceleration which we begin by stating that any object moving in a straight line will be accelerating if its velocity is increasing or decreasing during a given period of time. Acceleration can be either positive or negative depending on whether the velocity is increasing or decreasing. Thus, the acceleration is described as the rate of change of velocity of an object. Acceleration is a vector quantity that is described as the frequency at which the object's velocity changes. So, we can say in simple terms that

$$\text{Linear acceleration} = \frac{\text{Change in velocity}}{\text{Time taken}} \qquad (22.4)$$

and its unit is meter per second squared or m/s².

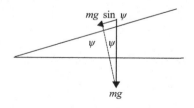

FIGURE 22.3 As shown here, hill climbing force is equal to $mg \sin \psi$.

Angular Acceleration

Angular acceleration refers to the time rate of change of angular velocity and is measured in units of angle per unit time square (which in SI units is radians per second squared) and is usually represented by the symbol alpha (α).

Tractive Force

Tractive force is simply the sum of all the forces we have discussed—it is the propulsion unit from an electric vehicle-on-wheels. As used in this text, the term tractive force refers to total traction a vehicle exerts on a surface.

NOTES

1 This chapter is based on Spellman, F. (2023).
2 Much of the information in this section is adapted from NASA (2017).

REFERENCES

BMW M3 E46 (2022). Car.info. Accessed 05/24/22 @ https://www.bmusa.com.
Car Handling Basics (2022). How to and design tips. Accessed 05/24/22 @ https://www. buildyourownracecar.com/race-car-handling-basics-and-design/.
Cortez, P. (2014). Repairing persistent pulls, drifts, shimmies & vibration. Accessed 05/28/22 @ https://classic.artsautomotive.com/GSP9700.htm.
NASA (2017). Aerodynamics. Accessed 06/01/22 @ https://www.nasa.gov/offices.
Norris, W. (1906). Steering. Accessed 05/24/22 @ https://archive.org/details/modernsteamroadw 00norrich.
Pacejka, H.B. (2005). *Tyre and vehicle dynamics* (2nd ed.). Warrendale, PA, SAE International, p. 22. Accessed 05/28/22 @ https://www.SAE.org.
Plasterk, K.J. (1989). The end of the first era: A farewell to Hans Pacejka. Accessed 05/27/2022 @ https://www.woldcat.org/issn/oo42-3114.
Spellman, F. (2023). *The science of electric vehicles*. Boca Raton, FL, CRC Press.

23 Electric Motors

INTRODUCTION

In this chapter, the discussion is about electric motors and their application in providing traction for electric vehicles (EVs). Oh! One might say. And one might ask: In a book about lithium, why is a basic discussion about motors important or included?

Really a good question with an easy-to-state answer. Remember that the lithium-ion battery used to power EVs actually powers the drive train, the traction wheel or traction wheels of the vehicle. It is the motor that drives the wheel or wheels. Therefore, it seems appropriate to discuss electric motors by providing a simplified discussion on and about electric motors. Note that the information provided in this chapter is based on material within *The Science of Electric Vehicles* by F. Spellman and published by CRC Press, Boca Raton, Florida.

MOTORS

A variety of different types of electric motors—and other important components—make up the power train that provides lithium-ion-battery-stored electrical energy that is converted to rotational power that is required to move an electric vehicle-on-wheels (see Figure 23.1). Of course, electric motors are used for other purposes than powering EVs. For instance, at least 60% of the electrical power fed to a typical waterworks and/or wastewater treatment plant is consumed by electric motors. One thing is certain: There is an almost endless variety of tasks that electric motors perform in industry and for personal use. An *electric motor* is a machine used to change electrical energy to mechanical energy to do the work. (Note that a generator does just the opposite; that is, a generator changes mechanical energy to electrical energy.)

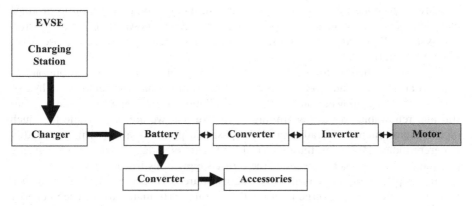

FIGURE 23.1 Electric motor and electric drive system components.

DOI: 10.1201/9781003387879-28

Previously, we pointed out that when a current passes through a wire, a magnetic field is produced around the wire. If this magnetic field passes through a stationary magnetic field, the fields either repel or attract, depending on their relative polarity. If both are positive or negative, they repel. If they are opposite polarity, they attract. Applying this basic information to motor design, an electromagnetic coil, the armature, rotates on a shaft. The armature and shaft assembly are called the rotor. The rotor is assembled between the poles of a permanent magnet and each end of the rotor coil (armature) is connected to a commutator also mounted on the shaft. A commutator is composed of copper segments insulated from the shaft and from each other by an insulting material. As like poles of the electromagnet in the rotating armature pass the stationary permanent magnet poles, they are repelled, continuing the motion. As the opposite poles near each other, they attract, continuing the motion.

As a quick review of importance of the electric motor in an EV, let's again point out that in an electric drive system, an electric motor converts the stored electrical energy in a battery to mechanical energy. Again, an electric motor consists of a rotor (the moving part of the motor) and a stator (the stationary part of the motor). A permanent magnet motor includes a rotor containing a series of magnets and a current-carrying stator (typically taking the form of an iron ring), separated by an air gap.

The technology involved with EVs is dynamic, constantly changing, advancing with the goal of EV manufacturers to improve the motors used to move the vehicles-on-wheels. Much of the current research and development (R&D) in progress is to improve motors in hybrid and plug-in electrical vehicles, with particular focus on improving batteries to increase vehicle range and also on reducing the use of rare earth materials currently used for permanent magnet-based motors.

With regard to rare earth elements (REEs), materials, and metals, they are critical to our modern way of life, although few people know or understand this. The truth be told, this lack of knowledge or understanding of REEs is surprising because they are critical ingredients in todays' mix of technologies including electronics, electric motors, magnets, batteries, generators, energy storage systems (supercapacitors/pseudocapacitors), emerging applications, and specialty alloys. REEs are used in various sectors of the U.S. economy including health care, transportation, power generation, petroleum refining, and consumer electronics.

Rare earth materials are extensively used in wind turbine operation to produce electrical power, in some solar applications for electrical power, and in energy storage systems. REEs are also used in EVs, thereby decreasing the need to use fossil fuels for operation.

Concern for the environment and for the impact of environmental pollution has brought about the trend (and the need) to shift from the use and reliance on hydrocarbons to energy-power sources that are pollution neutral or near pollution neutral and renewable. We are beginning to realize that we are responsible for much of the environmental degradation of the past and present—all of which is readily apparent today. Moreover, the impact of 200 years of industrialization and surging population growth has far exceeded the future supply of hydrocarbon power sources. So, the implementation of renewable energy sources is surging, and along with it there is a corresponding surge in utilization of rare earth materials for use in energy production.

So, the question is: Why are researchers trying to develop EV motors that do not utilize REEs. It is not because REEs are rare, because they are not. The problem is source(s); availability of REEs is basically controlled by countries outside the United States. Therefore, not only is the accessibility controlled by entities outside the United States but also the cost involved—economics is always in play in the manufacturing businesses. The United States possesses locations within this country to obtain REEs but with the exception of one location, REEs are not being actively mined.

DC MOTORS

The construction of a DC motor is essentially the same as that of a DC generator. However, it is important to remember that the DC generator converts mechanical energy into the electrical energy then back into mechanical energy. A DC generator may be made to function as a motor by applying a suitable source of DC voltage across the normal output electrical terminals. There are various types of DC motors, depending on the way the field coils are connected. Each has characteristics that are advantageous under given load conditions.

The field coils of *shunt motors* (see Figure 23.2) are connected in parallel with the armature circuit. This type of motor, with constant potential applied, develops variable torque at an essentially constant speed, even under changing load conditions. Such loads are found in machine-shop equipment such as lathes, shapes, drills, milling machines, and so forth.

The field coils of *series motors* (see Figure 23.3) are connected in series with the armature circuit. This type of motor, with constant potential applied, develops

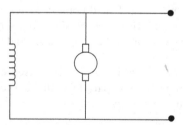

FIGURE 23.2   DC shunt motor.

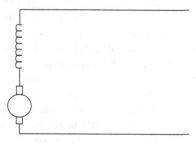

FIGURE 22.3   DC series motor.

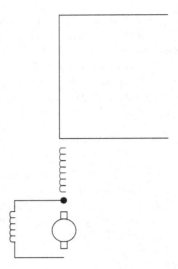

FIGURE 23.4 DC compound motor.

variable torque but its speed varies widely under changing load conditions. That is, the speed is low under heavy loads, but becomes excessively high under light loads. Series motors are commonly used to drive electric hoists, winches, cranes, and certain types of vehicles (e.g., electric trucks). In addition, series motors are used extensively to start internal combustion engines.

Compound motors (see Figure 23.4) have one set of field coils in parallel with the armature circuit, and another set of field coils in series with the armature circuit. This type of motor is a compromise between shunt and series motors. It develops an increased starting torque over that of the shunt motor and has less variation in speed than the series motor.

The speed of a DC motor is variable. It is increased or decreased by a rheostat connected in series with the field or in parallel with the rotor. Interchanging either the rotor or field winding connections reverses direction.

AC MOTORS

Alternating-current voltage can be easily transformed from low voltages to high voltages or vice versa and can be moved over a much greater distance without too much loss in efficiency. Most of the power generating systems today, therefore, produce AC. Thus, it logically follows that a great majority of the electrical motors utilized today are designed to operate on AC. However, there are other advantages in the use of AC motors besides the wide availability of AC power. In general, AC motors are less expensive than DC motors. Most types of AC motors do not employ brushes and commutators. This eliminates many problems of maintenance and wear and eliminates dangerous sparking. AC motors are manufactured in many different sizes, shapes, and ratings, for use on an even greater number of jobs. They are designed for use with either polyphase or single-phase power systems. This chapter cannot possibly cover all aspects of the subject of AC motors. Consequently, it will deal mainly with the operating principles of the two most common types—the *induction* and *synchronous motor.*

INDUCTION MOTORS

The induction motor is the most commonly used type of AC motor because of its simple, rugged construction and good operating characteristics. It consists of two parts: the stator (stationary part) and the rotor (rotating part). The most important type of polyphase induction motor is the *three-phase motor.*

Important note: A three-phase (3-θ) system is a combination of three single-phase (1-θ) systems. In a 3-θ balanced system, the power comes from an AC generator that produces three separate but equal voltages, each of which is out of phase with the other voltages by 120°. Although 1-θ circuits are widely used in electrical systems, most generation and distribution of AC current is 3-θ.

The driving torque of both DC and AC motors is derived from the reaction of current-carrying conductors in a magnetic field. In the DC motor, the magnetic field is stationary and the armature, with its current-carrying conductors, rotates. The current is supplied to the armature through a commutator and brushes. In *induction motors*, the rotor currents are supplied by electromagnet induction. The stator windings, connected to the AC supply, contain two or more out-of-time-phase currents, which produce corresponding mmfs. These mmfs establish a rotating magnetic field across the air gap. This magnetic field rotates continuously at constant speed regardless of the load on the motor. The stator winding corresponds to the armature winding of a DC motor or to the primary winding of a transformer. The rotor is not connected electrically to the power supply.

The induction motor derives its name from the fact that mutual induction (or transformer action) takes place between the stator and the rotor under operating conditions. The magnetic revolving field produced by the stator cuts across the rotor conductors, inducing a voltage in the conductors. This induced voltage causes rotor current to flow. Hence, motor torque is developed by the interaction of the rotor current and the magnetic revolving field.

With regard to induction motors used in various electric vehicles-on-wheels models, the high selling point, significant characteristic is that they have high starting torque and offer high reliability. Note, however, that induction motors have levels of power density and overall efficiency when compared to internal permanent magnetic (IPM) motors (described later). Induction motors are widely available and common in various industries today, including some production vehicles. There is a current problem with induction type motors used in EVs, they are dated—meaning they have been around a long time (aka old hat) and because of their maturity it is unlikely that research could achieve additional (new) improvements in cost, weight, efficiency, and volume for competitive future EVs. Simply, we know induction motors well and that is that.

SYNCHRONOUS MOTORS

Okay, let's get back to the basics (that is, to the foundation blocks of basic electric motor principles), we begin with synchronous motors. Like induction motors, *synchronous motors* have stator windings that produce a rotating magnetic field. However, unlike the induction motor, the synchronous motor requires a separate source of DC from the field. It also requires special starting components. These include a salient-pole field with starting grid winding. The rotor of the conventional type synchronous motor is essentially the same as that of the salient-pole AC generator. The stator windings of induction and synchronous motors are essentially the same.

In operation, the synchronous motor rotor locks into step with the rotating magnetic field and rotates at the same speed. If the rotor is pulled out of step with the rotating stator field, no torque is developed and the motor stops. Since a synchronous motor develops torque only when running at synchronous speed, it is not self-starting and hence needs some device to bring the rotor to synchronous speed. For example, a synchronous motor may be started rotating with a DC motor on a common shaft. After the motor is brought to synchronous speed, AC current is applied to the stator windings. The DC starting motor now acts as a DC generator, which supplies DC field excitation for the rotor. The load then can be coupled to the motor.

SINGLE-PHASE MOTORS

Single-phase (1-θ) motors are so called because their field windings are connected directly to a single-phase source. These motors are used extensively in fractional horsepower sizes in commercial and domestic applications. The advantages of using single-phase motors in small are that they are less expensive to manufacture than other types, and they eliminate the need for three-phase AC lines. Single-phase motors are used in fans, refrigerators, portable drills, grinders, and so forth.

A single-phase induction motor with only one stator winding and a cage rotor is like a three-phase induction motor with a cage rotor except that the single-phase motor has no magnetic revolving field at start and hence no starting torque. However, if the rotor is brought up to speed by external means, the induced currents in the rotor will cooperate with the stator currents to produce a revolving field, which causes the rotor to continue to run in the direction, which it was started.

Several methods are used to provide the single-phase induction motor with starting torque. These methods identify the motor as *split-phase, capacitor, shaded-pole*, and *repulsion-start* induction motor. Another class of single-phase motors is *the AC series* (universal) type. Only the more commonly used types of single-phase motors are described.

SPLIT-PHASE MOTORS

The split-phase motor (see Figure 23.5) has a stator composed of slotted lamination that contain a starting winding and a running winding.

Note: If two stator windings of unequal impedance are spaced 90 electrical degrees apart but connected in parallel to a single-phase source, the field produced will appear to rotate. This is the principle of *phase splitting*.

The starting winding has fewer turns and smaller wire than the running winding, hence has higher resistance and less reactance. The main winding occupies the lower

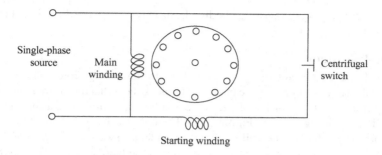

FIGURE 23.5   Split-phase motor.

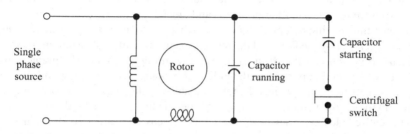

FIGURE 23.6   Capacitor motor.

half of the slots and the starting winding occupies the upper half. When the same voltage is applied to both windings, the current in the main winding lags behind the current in the starting winding. The angle θ between the main and starting windings is enough phase difference to provide a weak rotating magnetic field to produce a starting torque. When the motor reaches a predetermined speed, usually 75% of synchronous speed, a centrifugal switch mounted on the motor shaft opens, thereby disconnecting the starting winding. Because it has a low starting torque, fractional horsepower split-phase motors are used in a variety of equipment such as washers, oil burners, ventilating fans, and woodworking machines. Interchanging the starting winding leads can reverse the direction of rotation of the split-phase motor.

CAPACITOR MOTORS

The capacitor motor is a modified form of split-phase motor, having a capacitor in series with the starting winding. The capacitor motor operates with an auxiliary winding and series capacitor permanently connected to the line (see Figure 23.6). The capacitance in series may be of one value for starting and another value for running. As the motor approaches synchronous speed, the centrifugal switch disconnects one section of the capacitor. If the starting winding is cut out after the motor has increased in speed, the motor is called a *capacitor-start motor*. If the starting winding and capacitor are designed to be left in the circuit continuously, the motor is called *capacitor-run motor*. Capacitor motors are used to drive grinders, drill presses, refrigerator compressors, and other loads that require relatively high starting torque. Interchanging the starting winding leads may reverse the direction of rotation of the capacitor motor.

Shaded-Pole Motor

A shaded-pole motor employs a salient-pole stator and a cage rotor. The projecting poles on the stator resemble those of DC machines except that the entire magnetic circuit is laminated and a portion of each pole is split to accommodate a short-circuited coil called a *shading coil* (see Figure 23.7). The coil is usually a single band or strap of copper. The effect of the coil is to produce a small sweeping motion of the field flux from one side of the pole piece to the other as the field pulsates. This slight shift in the magnetic field produces a small starting torque. Thus, shaded-pole motors are self-starting. This motor is generally manufactured in very small sizes, up to 1/20 hp, for driving small fans, small appliances, and clocks.

In operation, during that part of the cycle when the main pole flux is increasing, the shading coil is cut by the flux, and the resulting induced emf and current in the shading coil tend to prevent the flux from rising readily through it. Thus, the greater portion of the flux rises in that portion of the pole that is not near the shading coil. When the flux reaches its maximum value, the rate of change of flux is zero, and the voltage and current in the shading coil are zero. At this time, the flux is distributed more uniformly over the entire pole face. Then as the main flux decreases toward zero, the induced voltage and current in the shading coil reverse their polarity, and the resulting mmf tends to prevent the flux from collapsing through the iron in the region of the shading coil. The result is that the main flux first rises in the unshaded portion of the pole and later in the shaded portion. This action is equivalent to a sweeping movement of the field across the pole face in the direction of the shaded pole. This moving field cuts the rotor conductors and the force exerted on them causes the rotor to turn in the direction of the sweeping field. The shaded-pole method of starting is used in very small motors, up to about 1/25 hp, for driving small fans, small appliances, and clocks.

REPULSION-START MOTOR

Like a DC motor, the *repulsion-start motor* has a form-wound rotor with commutator and brushes. The stator is laminated and contains a distributed single-phase winding.

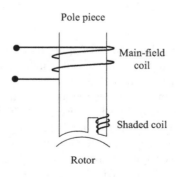

FIGURE 23.7 Shaded pole.

In its simplest form, the stator resembles that of the single-phase motor. In addition, the motor has a centrifugal device, which removes the brushes from the commutator and places a short-circuiting ring around the commutator. This action occurs at about 75% of synchronous speed. Thereafter, the motor operates with the characteristics of the single-phase induction motor. This type of motor is made in sizes ranging from 1/2 to 15 hp and is used in applications requiring a high starting torque.

SERIES MOTOR

The AC series motor will operate on either AC or DC circuits. When an ordinary DC series motor is connected to an AC supply, the current drawn by the motor is low due to the high series-field impedance. The result is low running torque. To reduce the field reactance to a minimum, AC series motors are built with as few turns as possible. Armature reaction is overcome by using *compensating windings* (see Figure 23.8) in the pole pieces. As with DC series motors, in an AC series motor the speed increases to a high value with a decrease in load. The torque is high for high armature currents so that the motor has a good starting torque. AC series motors operate more efficiently at low frequencies. Fractional horsepower AC series motors are called *universal motors*. They do not have compensating windings. They are used extensively to operate fans and portable tools, such as drills, grinders, and saws.

IPM MOTORS

Notwithstanding the IPM motors and relatively expensive it is their high-power density and maintenance of high efficacy over a high percentage of their operating range that makes the use of IPMs currently the motor of choice of a few EV manufacturers. The high costs are a result of the high prices of magnets and rotor fabrication—both these factors are expensive. Currently, almost all hybrid and plug-in electric vehicles-on-wheels use rare earth permanent magnets and this is a significant part of the expense. The rare earth magnets are expensive due to their limited availability. Even though the cost of rare earth magnets used in IPM motors appears to be increasing in price (as with almost all other products or services) EV manufacturers are likely to use these magnets until something more practical and less expensive is found as replacement.

SWITCHED RELUCTANCE MOTORS (SRM)

Because of its simple and rugged construction the SRM has been gaining interest in industrial applications such as renewable energy systems like wind power and

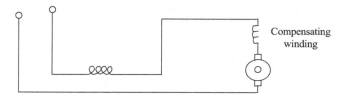

FIGURE 23.8   AC series motor.

in electric vehicles-on-wheels because it offers a lower cost option that is relatively easy to manufacture. Their rugged structure tolerates high temperatures and speeds. Well, with the good comes that not good: these motors produce more noise and vibration than comparable motor designs, which is a major challenge for use in vehicles-on-wheels. Another issue is that switched reluctance motors are less efficient than other motor types and required additional sensors and complex motor controllers that increase the overall cost of the electric drive systems.

The bottom line: No matter what electrical source of power is used to propel the EVs, there must be some type of an electric motor involved.

24 DC/DC Converters

INTRODUCTION

Figure 24.1 shows and highlights DC/DC converters (aka buck-boost converters) used in the electric drive system for many types of electric vehicles-on-wheels. DC/DC converters in an electrical vehicle may be classified as unidirectional (operates with various onboard loads) and bidirectional converters (used in regenerative backup, backup power, and recharging operations). DC/DC converters are used to increase (boost) or decrease (buck) battery voltages (typically 200 to 450 V) to accommodate the voltage needs of motors and other vehicles and components. If the vehicle's electric motor requires higher voltage, such as an internal permanent magnet motor, it will require a boost DC/DC converter. If a component requires lower voltage, such as most vehicle systems (lighting, entertainment, and other accessories), it will require a buck DC/DC converter that reduces the voltage to the 12–42 V level. Research on developing improved converters is active and on-going. The goal is to increase efficiency, miniaturize or reduce part count, and enable modular (i.e., modular in that it or they contain several parts that serve small functions but combine to serve purpose of the devices and modules can be removed and replaced), scalable devices (i.e., scalable in the sense that it allows for the capacity to accommodate a greater amount of usage).

TERMINOLOGY AND DEFINITIONS

Step-down (buck converter)—a converter where the output voltage is lower than the input voltage (such as a buck converter).

Step-up (boost converter)—a converter that outputs a voltage higher than the input voltage (such as a boost converter).

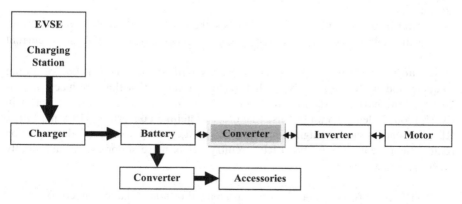

FIGURE 24.1 Electric drive system components highlighting DC/DC converters.

DOI: 10.1201/9781003387879-29

Coil-integrated DC/DC converters—decreases mounting space with a small number of components (power control IC, coil, capacity, and resistor) in a single integration solution.

Continuous current mode (CCM)—current and the magnetic field in the inductive energy storage never reaches zero—the current fluctuates but never goes down to zero.

Discontinuous current mode (DCM)—current and the magnetic field in the inductive energy storage may reach or cross zero—the current fluctuates during the cycle, going down to zero at or before the end of each cycle.

Galvanic isolation—this is a technique that separates electrical circuits to eliminate stray currents; no direct conduction path is permitted.

Hard switched—in topology refers to transistors switch quickly while expose to both full voltage and full current.

Input noise—if the converter loads the input with sharp load edges, the converter can emit RF noise from the supplying power lines. This is usually prevented by using proper filtering in the input stage of the converter.

Mainstream converter topologies—non-isolated types include boost, buck-boost, and buck; isolated types include flyback, forward, push-pull, half bridge, and full bridge. Other topologies with numerous variations include SEPIC (single-ended primary-inductor converter), Cuk converter, Current Fed Buck, Tapped Inductors, Multiple Outputs, and Interleaving.

Noise—unwanted electrical and electromagnetic signal noise, typically switching artifacts.

Output noise—even though an ideal DC-to-DC converter produces an output that is a flat, with constant output voltage, these converters produce DC output upon which is superimposed some level of electrical noise. Note that switching converters produce switching noise at the switching frequency and its harmonics. Moreover, all electronic circuits have some thermal noise (i.e., noise made by thermal agitation of the electrons flowing in the circuit). Because some electrical/electronic radio-frequency and analog circuits require a power supply with little noise, the use of linear regulators which work to maintain a constant voltage output and basically acts like a variable resistor, continuously adjusting to maintain an average value of output results.

Resonant—is an LC circuit that shapes the voltage across the transistor and current through it so that the transistor switches when either the voltage or current is zero.

RF noise—switching converters inherently emit radio waves at the switching frequent and its harmonics. Note that switching converters that produce triangular switching currents, such as the Split-Pi, forward converter, or Cuk converter in CCM, produce less harmonic noise that other switching converters (Hoskins, 1997). RF noise causes electromagnet interference (EMI). Acceptable levels depend upon requirements, e.g., proximity to RF circuitry needs more suppression than simply meeting regulations.

DC/DC converters can be classified as isolated or non-isolated converters.
Isolated or non-isolated converters?

Yes.

What is the difference?

Okay, good question. The difference is the non-isolated DC/DC converter that has a single switch and a single diode and also may have inductors and capacitors to store energy and are typically of the boost, buck-boost, or buck converters. Isolated DC/DC converters are derived from the basic topologies defined earlier and also include flyback, forward, push-pull, half bridge, and full bridge topologies.

Non-Isolated and Isolated Converter Topologies (aka Configurations)

Non-Isolated Converters

- **Boost converter (non-isolated)**—is a step-up converter that raises voltage while lowering current from its supply to its load.
- **Buck-boost converter**—is a type of switched mode power supply that combines the principles of the buck converter and boost converter in a single circuit and provides a regulated DV voltage from either an AC or a DC input.
- **Buck converter** (aka step-down converter) this DC-to-DC power concrete steps down voltage (while drawing less average current) from its supply (input) to its load (output). As a switched-mode-power-supply (SMPS)—often used to convert voltage in computers—it typically contains at least two semiconductors (a diode and a transistor); at present, the trend is to replace the diode with a second transistor and at least one energy storage element, a capacitor, inductor, or a combination of both.

Isolated Converters

- **Flyback converter**—is used in both AC/DC and DC/DC conversion with what is known as *galvanic isolation* between the input and any outputs. This is a buck-boost converter with the inductor split to form a transformer; this facilitates multiplied voltage ratios giving an additional advantage of isolation.
- **Forward converter**—this DC/DC converter uses a transformer to increase or decrease the output voltage based on the transformer's ratio and provides galvanic isolation for the load. This type of converter has multiple output windings making it possible to provide both higher and lower voltage outputs simultaneously.
- **Push-pull converter**—this converter is a type of DC/DC converter, basically a switching converter that uses a transformer to change the voltage of a DC power supply.
- **Half-bridge converter**—like flyback and forward converters, the half-bridge converter is type of converter that can supply an output voltage either higher or lower than the input voltage and provide electrical isolation via transformer.
- **Full-bridge converter**—this is a DC-to-DC converter configuration that employs four active switching components in a bridge configuration across

a power transformer. A full-bridge converter, besides reversing the polarity and providing multiple output voltages simultaneously also, is one of the commonly used configurations that offer isolation in addition to stepping up or down the input voltage.

REFERENCE

Hoskins, K. (1997). Making −5V Quiet, section of Linear technology application note 84. Accessed 6/12/22 @ http://www.linear.com/docs/4173.

25 Inverters

ELECTRIC VEHICLE INVERTERS

In Figure 25.1, an inverter is shown in the electric vehicle-on-wheels power train configuration. Not only is the inverter included in the power train of electric vehicles-on-wheels but it also is a principal component. The inverter shown in Figure 25.1 is needed to convert DC energy from a battery (remember batteries produce DC voltage and current) to AC power to drive the motor. An inverter also acts as a motor controller and as a filter to isolate the battery from potential damage from stray currents. In addition, because batteries produce DC voltage and current and most consumer products work on AC, including some electric vehicles-on-wheels, a vehicle power invertor is necessary to use the AC devices on the road.

Note that there is nothing new about using inverters in vehicles-on-wheels and other types of vehicles. At the present time, inverters are used in firefighting platforms and ambulances, recreational vehicles, buses, and in non-vehicles-on-wheels such as boats. With the increase in production of electrified vehicles-on-wheels, new and potentially stronger demand for inverters is currently occurring—with significant increased use for this purpose in the future.

Conventional vehicle inverters operate off of low voltage and are limited to low power (limits of battery and alternation supply) at 12–24 volts DC (VDC) and less than 6 kilowatts (kW). What we call "electric vehicles-on-wheels" in this text (and maybe elsewhere) battery electric, hybrid electric, and other fuel cell applications operate at higher DC (keep this in mind) voltage ranging from 48 to 800 VDC; in the kilowatt range, these vehicles put out 10–200 kW.

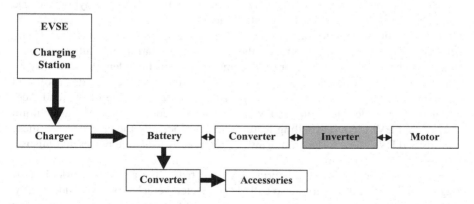

FIGURE 25.1 Vehicle-on-wheels electric drive system components highlighting inverter.

DOI: 10.1201/9781003387879-30

DID YOU KNOW?

Lithium batteries can be recharged 100's of times—a major advantage for this type battery.

ELECTRIC VEHICLES—PRESENT AND FUTURE

Currently, the push is on to switch from hydrocarbon vehicles-on-wheels to electric power-driven vehicles. There are numerous motives behind this movement with global climate change, the principal causal trend factor. The truth be told there is a movement toward electric vehicles-on-wheels that is presently gaining headway in the vehicle-on-wheels market. With gas prices averaging over $5.00/gallon (and seemingly rising), it is not surprising that electric vehicle-on-wheels have become the focus for possibly their purchase to desire of one of these vehicles. Gas prices and gas availability have at least made some potential vehicles-on-wheels customers take a look at the electrical vehicles as a possible substitute for hydro-driven powered vehicles.

So, you might think that the manufacturers of electric vehicles-on-wheels rely solely on the high cost of gasoline and diesel as the only selling point needed for their product. This is not the case, however. Competition between manufacturers of vehicles-on-wheels is a significant driver that requires manufacturers and sellers to provide "incentives" or "extras" to attract the buyer's attention and possibly earn a final sale.

Okay, what are the incentives, the extras that are magnets of attention for potential buyers of electric vehicles-on-wheels?

For some vehicles, especially for the on-road vehicles, it is the potential to have included their vehicle in 110 VAC outlets available for standard electrical applications. In addition, the special features of hybrid electric "contractor" trucks are attractive for those who need specialized trucks for their work. A hybrid electric contractor truck has a service body with a ladder rack. Important to a contractor is to have a quality vehicle to transport equipment and assist in completing jobs. Using this type of truck gives the contractor the ability and versatility to haul almost anything. A hybrid electric contractor truck comes in various body lengths with varying features and is basically equipped with the perfect body on the job site. Tools and accessories are securely stored and easily accessible. In the case of off-road electric hybrids and electric vehicles (EVs), the military, construction, and agriculture hybrids and EVs are a real selling feature, this is especially the case because these hybrids and EVs allow for operation away from readily available power and this is a very strong selling feature.

So, it looks like the future of hybrids, EVs, and trucks has a glowing outlook. This bright future also extends to the so called "connected vehicles" (cars). Additionally, high-capacity inverters are a big attraction and sell to the military, agriculture, and construction. Also, fuel cells of various sorts are likely to be innovative technologies that will impact transportation needs.

Note: Fuel cells and various innovations are addressed later in this book.

Well, all of this to this point is positive and encouraging in the transfer from hydrocarbon-powered vehicles to EVs and fuel-cell-powered vehicles on wheels. With regard to vehicle's electrical and mechanical characteristics, it can be said that electric vehicles generally have low sensitivity to perturbation. EVs, in most cases, are stand-alone vehicles-on-wheels. With regard to EVs' dimensions, their size and weight are important and should be small and lightweight. Reliability and component life requirements are important and the goal should be to ensure 10-year, 150,000-mile, 5,000-hour vehicle component life.

The bottom line: cost is the overriding figure of merit.

There is more to consider too. Today's EV inverters are missing, lacking, or short of low cost—cost is a huge factor. Also, for a given output, the inverters need to be smaller in size. They also need to be constructed with built-in ability to handle vibration, shock, and extreme temperatures. Inverters must be designed to provide higher power capability and unfortunately at the present time, there are few inverter types available for this burgeoning market.

There are important issues related to inverter usage and availability. For example, performance is an issue. High output is required but in a small package, able to survive and function properly in difficult environments. Also, the functionality built into some existing inverter produces is often not required.

And then there is the cost. The truth be told electric vehicles-on-wheels, hybrid electric, and fuel cell vehicles are too expensive for the average person. However, with regard to market share, the market for advanced vehicle inverters is wide open (basically unprotected) to competitors. However, many of the competitors have tired of waiting for the market to develop and have dropped out; at least for the time being.

Part 6

Fuel Cells

26 Fuel Cells

WHAT IF?

What if a traveler could jump into a vehicle-on-wheels and drive 320 miles before the vehicle's battery needed to be recharged? What if a traveler could travel 320 miles without adding pollution to the atmosphere? What if the traveler could travel 320 with very little, if any, vehicle-on-wheels created noise?

Okay, what if you were driving a vehicle-on-wheels powered by a fuel cell instead of a lithium-battery? What if you were able to drive a fuel cell vehicle-on-wheels that could give you more mileage than an electric vehicle? What if your fuel cell vehicle-on-wheels produces zero pollutants as you drive along?

What if any of the above statements about the benefits of electric and fuel-cell–powered vehicles-on-wheels is reality?

Well, the truth be told, the above statements about the benefits of electric and fuel cell vehicles on wheels are reality—in a limited sense, of course.

The problem is that when there are benefits derived from anything, there is or are the opposite results or occurrences; that is, there is/are drawback(s).

Let's look at the problem(s).

First, *range anxiety* is the main driver of concern, apprehension, angst, and fear of the driver of the electric or fuel-cell–powered vehicles-on-wheels.

So, what is range anxiety?

Simply stated, range anxiety is a driver's fear that a vehicle-on-wheels has insufficient energy storage (battery capacity and/or fuel cell capacity) to cover the highway distance needed to reach its intended destination and therefore strands the vehicle-on-wheels occupants somewhere along the highway, far from their intended destination (Backstrom, 2009; Schott, 2009; Eberle and von Helmolt, 2010; Rahim, 2010). Range anxiety applies to battery electric vehicles (BEVs) and fuel-cell–powered vehicles. Range anxiety is considered to be one of the major psychological barriers to large-scale public adoption of electric and fuel cell cars (Eberle and von Helmolt, 2010; Gomez-Ibane et al., 2010; Gordon-Bloomfield, 2001).

The problem with electric and fuel-cell–powered vehicles and the main contributor to driver's range anxiety is exacerbated by the present lack of infrastructure to provide battery charging and also fuel cell changeout. It is basically concern about drivers becoming stranded. Running out of gasoline or diesel is not fun; however, one can obtain a couple gallons of these hydrocarbon fuels rather easily by hitching a ride to or walking to nearest gas station and obtaining fuel to get one's vehicle back up and running to the nearest fill-up station.

The solution to range anxiety and other power source problems is swap-out—swapping the battery. At the change out station, the depleted battery is swapped for fully charge battery in about the time takes to fill one's gas tank.

Sounds like the answer to reducing range anxiety, doesn't it?

DOI: 10.1201/9781003387879-32

Well, not exactly. The manner in which vehicle-on-wheels' batteries are mounted on the undercarriage of the vehicle makes it in no way an easy swap-out with a fully changed battery. This can change, of course, with the technological advancements occurring daily, it may be possible to make the battery change out—in the future.

For now, it is more likely, potentially possible that the fuel cell is the power package that is easiest to change out at the swapping station. This could be the case, could be enhanced or developed if the hydrogen or some other fuel cell energy source ingredient(s) could be packaged into a removable cylinder that is about 4 in diameter and about 4 ft in length and weighing no more than 10 pounds (lbs). And it is for this reason, the ability to swap out the fuel cell at a swapping stations (when this station is one of thousands throughout the United States.) that can provide a full energy cell with relative ease that is so attractive about this type of power source.

So, while the hydrogen fuel cell and other types of fuel cells are possibly the answer to shifting from hydrocarbon fuels, it is important to discuss them even though they are not lithium powered. Thus, this chapter is provided to show the fuel cell and how it compares to any type of battery-driven vehicle-on-wheels. Keep in mind that there is no present budget hydrogen vehicles and that the cost of change out fuel cell cylinders is or might be cost prohibitive. But fuel cells may be the answer to: What if?

FUEL CELLS: THE BASICS

In regard to the term "cell," depending on your education level, social, cultural, and/or economic background, this term may conjure up initial meanings that are as diverse in variety as are the colors, sizes, and shapes of lightning bolts (by the way, a huge source of "renewable" energy). For instance, some may think of the term cell as referring in terms related to plant cell, animal cell, cell structure, cell biology, cell diagram, cell membrane, human memory, cell theory, cell wall, cell parts, cell function, honeycomb cell, prison cell, electrolytic cell (for producing electrolysis) aeronautic gas cell (contained in a balloon), ecclesiastical cell (i.e., as used to describe a monastery or nunnery), or currently and definitely more commonly used to describe a cell phone.

You may have noticed that nowhere in this particular list are terms related to one of the topics of this book: lithium and in particular lithium-ion batteries. While some would argue that any kind of biological cell mechanism and/or part/organelle and cell phones are types of energy devices, producers or consumers that can be renewed unless destroyed, it is not our intention to argue this point or issue either way; instead, it is our point that an important type of cell is not included in the above list—and that it should be included. We are referring to the electric cell, electrochemical cell, galvanic cell, and voltaic cell—often referred to as a battery. This type of cell, no matter what we call, it is a device that generates electrical energy from chemical energy, usually consisting of two different conducting substances placed in an electrolyte. The bottom line on this type of electric cell is that it is similar to other unmentioned cells. Fuel cells.

Before moving on to our basic discussion of fuel cells and their associated terminology and applications, it is important to point out that although fuel cells are not the topics of discussion anywhere near as common as those other cells (cell phones), for instance, we predict that the day is coming when we will refer to our fuel cells just as commonly as we mention our cell phones—for it will be the fuel cell that will power our lives, just as the cell phone powers and transmits our communication.

OPEN CELLS VS CLOSED CELLS

Batteries store electrical energy chemically (contrary to popular belief, they do not "make" electrical energy); hence, they are thermodynamically open systems. By contrast, fuel cells are different form conventional electrochemical cell batteries in that they consume reactant from an external source, which must be replenished (Science Reference Services, 2007)—a thermodynamically open system.

A fuel cell is an electrochemical cell that converts a source fuel (hydrogen gas or a hydrogen-rich liquid fuel can also use hydrocarbons, alcohols, chlorine and chlorine dioxide, and others) into an electrical current. The hydrogen gas or hydrogen rich-fuel is reacted or triggered in the presence of an electrolyte (an oxidant), usually oxygen from the air, to produce electrical, heat, and water. The reactants flow into the cell and the reaction products flow out of it, while the electrolyte remains with it. Fuel cells can operate continuously as long as the reactant and oxidant flows are maintained.

Keep in mind that fuel cells come in many varieties (see Table 26.1); however, they all work in the same general manner; the principle of operation of a hydrogen-type fuel cell is described in the next section.

TABLE 26.1
Types of Fuel Cells

Fuel Cell Type	Electrolyte	Current Use(s)
Metal hydride	Aqueous alkaline solution	Commercial/research
Electro-galvanic	Aqueous alkaline solution	Commercial/research
Direct formic acid	Polymer membrane (ionomer)	Commercial/research
Zinc-air battery	Aqueous alkaline solution	Mass production
Microbial	Polymer membrane/humic acid	Research
Upflow microbial	*****	Research
Regenerative	Polymer membrane (ionomer)	Commercial/Research
Direct borohydride	Aqueous alkaline solution	Commercial/Research
Alkaline	Aqueous alkaline solution	Commercial/research
Direct methanol	Polymer membrane (ionomer)	Research
Reformed methanol	Polymer membrane (ionomer)	Commercial/research
Direct-ethanol	Polymer membrane (ionomer)	Research
Proton exchange membrane	Polymer membrane (ionomer)	Commercial/research
RFC-Redox	Liquid electrolytes w/redox Shuttle and polymer membrane (ionomer)	Research
Phosphoric acid	Molten phosphoric acid (H_3PO_4)	Commercial/research
Molten carbonate	Molten alkaline carbonate	Commercial/research
Tubular solid oxide	O^2—conducting ceramic oxide	Commercial/research
Protonic ceramic	H^+—conducting ceramic oxide	Research
Direct carbon	Several different	Commercial/Research
Planar solid oxide	O^2—conducting ceramic oxide	Commercial/Research
Enzymatic biofuel	Any that will not denature the Enzyme	Research
Magnesium air	Salt water	Commercial/research

FUEL CELLS: A REALISTIC VIEW

HYDROGEN[1]

Containing only one electron and one proton, hydrogen, chemical symbol H, is the simplest element on the Earth. Hydrogen is a diatomic molecule—each molecule has two atoms of hydrogen (which is why pure hydrogen is commonly expressed as H_2). Although abundant on the Earth as an element, hydrogen combines readily with other elements and is almost always found as part of another substance, such as water hydrocarbons, or alcohols. Hydrogen is also found in biomass, which includes all plants and animals.

- Hydrogen is an energy carrier, not an energy source. Hydrogen can store and deliver usable energy, but it doesn't typically exist by itself in nature; it must be produced from compounds that contain it.
- Hydrogen can be produced using diverse, domestic resources including nuclear; natural gas and coal; and biomass and other renewables including solar, wind, hydroelectric, or geothermal energy. This diversity of domestic energy sources makes hydrogen a promising energy carrier and important to our nation's energy security. It is expected and desirable for hydrogen to be produced using a variety or resources and process technologies (or pathways).
- DOE focuses on hydrogen-production technologies that result in near-zero, net greenhouse gas emissions and use renewable energy sources, nuclear energy, and coal (when combined with carbon sequestration). To ensure sufficient clean energy for our overall energy needs, energy efficiency is also important.
- Hydrogen can be produced via various process technologies, including thermal (natural gas reforming, renewable liquid and bio-oil processing, and biomass and coal gasification), electrolytic (water splitting using a variety of energy resources), and photolytic (splitting water using sunlight via biological and electrochemical materials).
- Hydrogen can be produced in large, central facilities (50–300 miles from point of use), smaller semicentral (located within 25–100 miles of use) and distributed (near or at point of use). Learn more about distributed vs. centralized production.
- In order for hydrogen to be successful in the market place, it must be cost-competitive with the available alternatives. In the light-duty vehicle transportation market, this competitive requirement means that hydrogen needs to be available untaxed at $2–$3/gge (gasoline gallon equivalent). This price would result in hydrogen fuel cell vehicles (FCVs) having the same cost to the consumer on a cost-per-mile-driven basis as a comparable conventional internal-combustion engine or hybrid vehicle.
- DOE is engaged in research and development of a variety of hydrogen production technologies. Some are further along in development than others—some can be cost-competitive for the transition period (beginning in 2015), and others are considered long-term technologies (cost-competitive after 2030).

Infrastructure is required to move hydrogen from the location where it's produced to the dispenser at a refueling station or stationary power site. Infrastructure includes the pipelines, trucks, railcars, ships, and barges that deliver fuel, as well as the facilities and equipment needed to load and unload them.

Delivery technology for hydrogen infrastructure is currently available commercially, and several U.S. companies deliver bulk hydrogen today. Some of the infrastructure is already in place because hydrogen has long been used in industrial applications, but it's not sufficient to support widespread consumer use of hydrogen as an energy carrier. Because hydrogen has a relatively low volumetric energy density, its transportation, storage, and final delivery to the point of use comprise a significant cost and result in some of the energy inefficiencies associated with using it as an energy carrier.

Options and trade-offs for hydrogen delivery from central, semicentral, and distributed production facilities to the point of use are complex. The choice of a hydrogen production strategy greatly affects the cost and method of delivery.

For example, larger, centralized facilities can produce hydrogen at relatively low costs due to economies of scale, but the delivery costs for centrally produced hydrogen are higher than the delivery costs for semicentral or distributed production options (because the point of use is farther away). In comparison, distributed production facilities have relatively low delivery costs, but the hydrogen production costs are likely to be higher—lower volume production means higher equipment costs on a per-unit-of-hydrogen basis.

Key challenges to hydrogen delivery include reducing delivery cost, increasing energy efficiency, maintaining hydrogen purity, and minimizing hydrogen leakage. Further research is needed to analyze the trade-offs between the hydrogen production options and the hydrogen delivery options taken together as a system. Building a national hydrogen delivery infrastructure is a big challenge. It will take time to develop and will likely include combinations of various technologies. Delivery infrastructure needs and resources will vary by region and type of market (e.g., urban, interstate, or rural). Infrastructure options will also evolve as the demand for hydrogen grows and as delivery technologies develop and improve.

Hydrogen Storage

Storing enough hydrogen on-board a vehicle to achieve a driving range of greater than 300 miles is a significant challenge. On a weight basis, hydrogen has nearly three times the energy content of gasoline (120 MJ/kg for hydrogen versus 44 MJ/kg for gasoline). However, on a volume basis, the situation is reversed (8 MJ/L for liquid hydrogen versus 32 MJ/L for gasoline). On-board hydrogen storage in the range of 5–13 kg H_2 is required to encompass the full platform of light-duty vehicles.

Hydrogen can be stored in a variety of ways, but for hydrogen to be a competitive fuel for vehicles, the hydrogen vehicle must be able to travel a comparable distance to conventional hydrocarbon-fueled vehicles.

Hydrogen can be physically stored as either a gas or a liquid. Storage as a gas typically requires high-pressure tanks (5,000–10,000 psi tank pressure). Storage of hydrogen as a liquid requires cryogenic temperatures because the boiling point of hydrogen at one atmosphere pressure –252.8°C.

Hydrogen can also be stored on the surfaces of solids (by adsorption) or within solids (by absorption). In adsorption, hydrogen is attached to the surface of material either as hydrogen molecules or as hydrogen atoms. In absorption, hydrogen is dissociated into H-atoms, and then the hydrogen atoms are incorporated into the solid lattice framework.

Hydrogen storage in solids may make it possible to store large quantities of hydrogen in smaller volumes at low pressures and at temperatures close to room temperature. It is also possible to achieve volumetric storage densities greater than liquid hydrogen because the hydrogen molecule is dissociated into atomic hydrogen within the metal hydride lattice structure.

Finally, hydrogen can be stored through the reaction of hydrogen-containing materials with water (or other compound such as alcohols). In this case, the hydrogen is effectively stored in both the material and in the water. The term "chemical hydrogen storage" or chemical hydrides is used to describe this form of hydrogen storage. It is also possible to store hydrogen in the chemical structures of liquids and solids.

Hydrogen Fuel Cell

As mentioned, the fuel cell uses the chemical energy of hydrogen to cleanly and efficiently produce electricity with water and heat as byproducts. Fuel cells are unique in terms of variety of their potential applications; they can provide energy for systems as large as a utility power station and as small as a laptop computer.

Fuel cells have several benefits over conventional combustion-based technologies currently used in many power plants and passenger vehicles. They produce much smaller quantities of greenhouse gases and none of the air pollutants that create smog and cause health problems. If pure hydrogen is used as a fuel, fuel cells emit only heat and water as byproducts.

DID YOU KNOW?

Hydrogen fuel cell vehicles (FCVs) emit approximately the same amount of water per mile as vehicles using gasoline-powered internal combustion engines (ICEs).

A hydrogen fuel cell is a device that uses hydrogen (or hydrogen-rich fuel) and oxygen to create electricity by an electrochemical process. A single fuel cell consists of an electrolyte and two catalyst-coated electrodes (a porous anode and cathode). Again, while there are different fuel cell types, all fuel cells work similarly:

- Hydrogen, or a hydrogen-rich fuel, is fed to the anode where a catalyst separates hydrogen's negatively charged electrons from positively charge ions (protons).
- At the cathode, oxygen combines with electrons and, in some cases, with species such as protons or water, resulting in water or hydroxide ions, respectively.

- For polymer electrolyte membrane and phosphoric acid fuel cells, protons move through the electrolyte to the cathode to combine with oxygen and electrons, producing water and heat.
- For alkaline, molten carbonate, and solid oxide fuel cells, negative ions travel through the electrolyte to the anode where they combine with hydrogen to generate water and electrons.
- The electrons from the anode cannot pass through the electrolyte to the positively charged cathode; they must travel around it via an electrical circuit to reach the other side of the cell. This movement of electrons is an electrical current.

NOTE

1 Information in this section from USDOE (2008).

REFERENCES AND RECOMMENDED READING

Backstrom, M. (2009). Comments of Southern California Edison Company on the California Public Utilities Commission Staff's White Paper, Light-Duty V. Accessed 12/25/22 @ http://www.cpuc.ca.gov/NR/rdonlyres/8B8F7624-1DF1-426B-847A-74255D3484D6/0/.

Eberle, U., and von Helmolt, R. (2010). Sustainable transportation based on electric vehicle concepts: A brief overview. Royal Society of Chemistry. Accessed 12/25/22 @ https://www.reserachgate.net/publication/224880220.

Gomez-lbane, J., Bennett, H.L., Haigh, D.J., Wiederer, A., and Philip, R. (2010). Policy options for electric vehicle charging infrastructure in C40 cities. Accessed 12/25/22 @ http://www.innovations.harvard.edu/cache/documents/11089/1108934.pdf.

Gordon-Bloomfield, N. (2001). Electric car out of juice? Pray for an Angel. Accessed 12/25/22 @ http://www.foxnews.com/leisure/2010/09/16/electric-car-juice-pray-angel/.

Rahim, S. (2010). Will lithium-air battery rescue electric car drivers from 'range anxiety'? Accessed 12/25/22 @ https://www.nytimes.com/cwire/2010/05/07/07climatewire-will-lithium-air-battery-rescue-electric-car-37498.html.

Schott, B. (2009). Range anxiety. Accessed 12/25/22 @ http://schott.blogs.nytimes.come/2009/01/15/range-ansxiety/.

Science Reference Services (2007). Batteries, supercapacitors, and fuel cells: Scope. Accessed 08/08/10 @ http://www.oc.gov/rr.scietech/tracer-bullets/batteriestb.html#scompe.

USDOE (2008). Hydrogen, fuel cells & infrastructure technologies program. Accessed @ http://www1.eere.energy.gov/hydrogenandfuelcells/production/basics.html.

Part 7

Final Word on Environmental and Safety Concerns

27 Environmental Concerns

INTRODUCTION

To this point, we have described how lithium is obtained from natural sources. What we have not discussed is that lithium is also derived from anthropogenic (human-caused) sources. As already noted, lithium is present in trace amounts in most rocks and natural waters and is a major component only in certain rare pegmatite minerals, clays, zeolites, and brines. Potential human-caused lithium sources include mine wastes, and after disposal, the wide range of manufactured products—part of what is commonly called the "throwaway society."

In soils, lithium is found in detrital (i.e., derived from transported rock fragments) and authigenic minerals (i.e., found where formed) and, to a lesser extent, in the organic fraction (Heier and Billings, 1978; Schrauzer, 2002). The concentration of lithium present is dependent upon the parent rock; a study of Scottish soils shows the highest lithium concentrations are found in soils that develop over clay stones and metamorphosed felsic rocks. Note that because of weathering and leaching, the top 10 cm of a soil profile can be poorer in lithium than underlaying clay minerals (Mitchell, 1964). Shacklette et al. (1977) observed that soils from the Western United States are more enriched in lithium than are those in the Eastern United States, with mean values of 23.3-ppm lithium and 17.3-ppm lithium, respectively. The lithium contents in sediments in hydrologically open and closed basins in California and Nevada are different (Cannon et al., 1975). In open basins, the lithium concentrations are relatively low, ranging from 0 to 150 ppm. In contrast, the range of lithium in hydrologically closed basins is 30–2000 ppm, and lacustrine clays have a median value of 700-ppm lithium. Why the difference? It's all about water; more specifically, about the flow of water. Given enough time, nothing can prevent water from moving in streams, brooks, rivers, and groundwater aquifers. When water moves, it carries with it passengers, sometimes in massive quantities and assorted types that are difficult to count such as dissolved salts. Therefore, the hydrological flow patterns in open basins with outflowing waters carry off sediments and dissolved salts while whatever is contained in the closed basin remains put.

Note that although lithium is not required for plant growth, they are known to assimilate it, typically in the range of 0.2–30 ppm (Aral and Vecchio-Sadus, 2008). Also note that more extreme values have also been reported; for example, matrimony vine (*Lycium barbarum*) from Arizona contained 1,120-ppm lithium (dry weight) (Sievers and Canon, 1973). A study of plants from 20 basins in Nevada and California demonstrated that plants grown in arid basins absorb more lithium than those in more humid regions. Canon et al. (2020) reported median lithium concentrations of 22.8 ppm (dry weight) from *Magnoliopsida* (aka dicotyledons) and basins compared with 1.3 ppm from humid regions of the United States (Bertrand, 1952).

DOI: 10.1201/9781003387879-34

In another study, Cannon et al. (1975) reported that plants growing in hot-spring discharge zones appear to have died from accumulations and precipitation of evaporative salts; these "fossilized root crowns" of salt can contain 1,000-ppm lithium.

Because lithium is a highly soluble element, it is commonly found as dissolved species in both groundwater and surface water. This attribute makes lithium useful as a conservative trace in hydrogeological studies (Bencala et al., 1990). There is evidence, however, that lithium may be removed from solution through sorption onto suspended clays. The Saline Valley in California is a good example of this where 98% of lithium is lost from solution during surface flow from the spring to the playa (Lombardi, 1963). Note that lithium readily substitutes for magnesium in the clay structure because of the similar ionic radii of the two elements, and this isomorphic substitution is likely responsible for the affinity of lithium in clays. Lithium can also be removed from solution, especially soil solution, by plants. Research and observation have shown that there is a much greater uptake of lithium by plants in acid soils, as acidity increased the solubility of metals (Schrauzer, 2002; Lenntech, 2007).

The increasing demand for electric vehicles (EVs) is anticipated to intensify the need for and use of lithium batteries. This increasing demand for lithium is occurring in spite of recent press reports of explosions and fires associated with lithium batteries which cause concerns about recycling (Mouawad and Drew, 2013). Note, however, that totally spent lithium batteries contain virtually no lithium metal because it has all been concerted to lithium oxide by the time the battery stops working (NEMA, 2001). Unlike lithium metal, lithium oxide is unreactive, it does not produce hydrogen gas upon exposure to water, and it does not catch fire. Both Hazardous Materials Identification System (HMIS) and National Fire Protection Association (NFPA)—discussed in detail in the next chapter—give it a "0" rating for flammability and physical (reactivity) hazard risks (Alpha Aesar, 2012; Sigma-Aldrich, 2014). Keep in mind, however, that lithium oxide is a health hazard rate a "3" by HMIS and NFPA because it is corrosive and can cause serious eye and skin damage upon contact. Note that disposal of spent batteries in a modern municipal solid waste facility is not likely to present a lithium hazard. According to ESPI Metals (1993) is moderately soluble—a maximum of 6.67 g of Li_2O can dissolve in 100 g of water (H_2O) at 0°C.

MINE WASTE

The type of lithium deposit affects mine-waste features. Only two types of deposits—pegmatites and closed-base brines—have produced mine wastes. Note that the mining history of lithium pegmatites may affect the mineralogy of tailings and waste piles that remain. Tin was the original target of mining at Kings Mountain (Kesler, 1942, 1955, 1961), suggesting that lithium-bearing minerals may still be remaining dumps from that era. Mineral processing of the pegmatite minerals involved crushing, wet grinding, sieving, gravity concentration, flotation, and collection with fatty acid amines where the tailings were discharged to storage ponds (Garrett, 2004) that likely contain some lithium. To improve the recovery of lithium, mineral processing was changed in 1969. The new processed used involved heating (calcining) the spodumene concentrate after collection, then leaching the calcine with hot water. Even with this new process, however, all the lithium still was not recovered.

In the grinding and sieving steps of processing, lithium ends up in waste and tailing piles via dissolution if the lithium minerals. Spodumene, for example, is fairly insoluble in water and dilute acids, but a small amount of dissolution has been observed during ore processing and concentration (Aral and Vecchio-Sadus, 2008). The dissolved lithium will be transported with the processing water and may concentrate in tailing ponds that have concentrations ranging up to 13 ppm.

Various evaporite minerals are produced in the production of lithium by evaporation of closed-basin brines. Details vary among the deposits owing to differences in brine chemistry. Clayton Valley precipitates include clay, calcium carbonate, calcium sulfate, hydrated magnesium, potassium-sodium sulfate, and potassium-sodium chloride (Garrett, 2004).

HEALTH CONCERNS

Lithium is not considered one of the essential elements for life (as are oxygen, carbon, hydrogen, and others), but it is present in most organisms in trace quantities (Leonard et al., 1995; Lenntech, 2007). Note that the normal human intake is about 2 milligrams per day (mg/d) (Leonard et al., 1995), and the human body contains about 7 milligrams (mg). Ingested lithium is absorbed from the intestines and excreted through the kidneys, and the half-life in the body is about 24 hours (Aral and Vecchio-Sadus, 2008).

DID YOU KNOW?

As mentioned earlier lithium may be used to treat mania associated with bipolar disorder. However, note that experts are not exactly sure how lithium works but believe it alters sodium transport in nerve and muscle cells which adjust the metabolism of neurotransmitters within the cell. Simply stated, we do not know what we do not know about lithium.

Modern use of lithium compounds for the treatment of bipolar issues dates from pioneering work after World War II in Australia by John Cade (Mitchell and Hadzi-Pavlovic, 2000). There are some things and/or events and/or the histories, biographies of certain geniuses who have made significant contributions to treating ill patients back to good health they are worth repeating and they are done so here again—hard work by open-mindedness, deep thinking, far thinkers are worth repeating; again, they are here:

AGAIN, DID YOU KNOW?

Dr. John Cade fought for Australia during World War II. Although trained as a psychiatrist, he served as a military surgeon. He served in Singapore and after its fall he became a prisoner of war from September 1941 to September 1945. While he was imprisoned, he observed various forms of behavior by his fellow prisoners. Of the various forms of behavior Dr. Cade observed were some fellow inmates who not only acted strange but also exhibited vacillating behavior (i.e., some inmates wavered in mind and/or opinion). Dr. Cade surmised that is was a toxin of some type that was affecting their brains and actions and when it was eliminated through their urine, they lost their symptoms.

After the war, Dr. Cade took up a position at a hospital in Melbourne. In an unused kitchen, Dr. Cade conducted crude experiments which eventually led to the discovery of lithium as a treatment mania (aka bipolar disorder). After ingesting lithium himself to ensure its safety in humans (Cade, 1949), Dr. Cade conducted a trial of lithium citrate and/or lithium carbonate on some of his patients diagnosed with mania, melancholia, and dementia, with exceptional results. Because his observations made during his treatment of his patients were so strong, he speculated that the cause of mania is the result of lithium deficiency (Cade, 1949).

Note that about 4% of people—more than 3 million out of the present population—will suffer from bipolar disorder at some point in their lives (Ketter, 2010). The exact number of those who treat the condition with lithium is unknown but undoubtedly is in the millions. Normal dosage? Well, for example, in 2013, a typical dosage was 1,800 mg/d of lithium carbonate (Drugs.com, 2013). At the present time (2023), dosage of 1,800 mg/d lithium carbonate is typically administered for acute control and long-term control ranges from 900 to 1,200 mg/day.

It is important to recognize that the lithium can be toxic. Because of this, lithium for medicinal use was banned in the United States between 1949 and 1970 (Strobusch and Jefferson, 1980). The various harmful effects including notice that a dose of 5 g of lithium can be fatal to humans was summarized by Aral and Vecchio-Sadus (2008). In this same timeframe, Aral and Vecchio-Sadus pointed out that no lithium poisoning had been reported from industrial applications.

ENVIRONMENTAL IMPACT

In addition to lithium's effect on many organisms—e.g., it disturbs the development of invertebrates, and in rats, it has been reported to reduce the number in a litter and the weight of the offspring and to cause incomplete ossification (conformity), and lithium also affects metabolism, neuronal communication, and cell proliferation in humans (Leonard et al., 1005; Domingo, 1994). One of the most common endpoints used in toxicity testing is the concentration that leads to 50% mortality (lethal concentration 50, LC_{50}) after exposure to a substance for a certain amount of time. In a study of earthworms exposed to soil containing lithium, the LC_{50} value was 70 milligrams per kilogram (mg/kg) after 7 weeks of exposure (Fischer and Molnar, 1997). In another study, the sensitivity of fish to lithium was measured in a series of 96-hour tests, which revealed an LC_{50} value that ranged from 13 mg/L for fathead minnows to greater than 100 mg/L for other fish species. Lithium concentrations of 1.2 mg/L immobilized the tiny crustacean *Daphnia magna* (aka water flea) in 64 hours, and a dose of 1.7 mg/L of lithium prevented the formation of embryos in fish eggs (Kszos and Stewart, 2003).

ENVIRONMENTAL FOOTPRINT

Lithium's importance to the world's carbon budget derives from its use in lightweight, rechargeable EV batteries, which have a smaller environmental footprint than conventional internal combustion engine vehicles. There are those, however, who point out that in order to recharge and produce the EV battery, an energy source (e.g., coal) is required and can induce environmental impacts. Others argue that over the course of a hypothetical lifetime of 150,000 miles (240,000 km), the average gasoline-powered car will produce about 64,000 kg of carbon dioxide based on values from the U.S. Environmental Protection Agency, Office of Transportation and Air Quality (2011). This greenhouse gas will go directly from the tailpipe to the atmosphere, where it will contribute to global climate change; some of it will later end up dissolved in seawater, where it will contribute to further ocean acidification, harming or killing fish and shellfish. The carbon dioxide of an EV depends largely on the means of charging its battery, but in all cases—even if the electricity from a coal-fired power plant—the output of carbon dioxide is lower than for a gasoline-powered vehicle (Notter et al., 2010). In the best-case scenarios, where the battery is recharged from renewable energy sources such as hydroelectric, wind, tidal, or solar power, the carbon dioxide output (beyond investment and infrastructure costs) is negligible.

THE OTHER SIDE OF THE COIN

As with other critical elements and materials, lithium resources have not been as thoroughly scrutinized as some conventional metallic resources, such as lead and copper. Stepped up research on the better-known lithium deposit types (brines and pegmatites) as well as on the newer deposit types (clays and zeolites) will help us to know what we do not know about lithium.

The bottom line: What it really boils down to is that we need to find out how we can get the maximum potential energy out of lithium-derived power storage devices and ensure that we have access to an adequate supply of lithium. This is especially the case if we are to turn off the hydrocarbon drilling, processing, and refining the spickets.

REFERENCES

Alpha Aesar (2012). Lithium oxide: Alpha Aesar safety data sheet, stock number 41832, review date August 16, 2021, 4 p. Accessed 01/05/23 @ http://www.alfa.com/contend/msds/USA/42832 pdf.

Aral, H., and Vecchio-Sadus, A. (2008). Toxicity of lithium to humans and the environment—A literature review. *Ecotoxicology and Environmental Safety*, v. 70, no. 3, pp. 349–356.

Bencala, K.E., McKnight, D.M., and Zellweger, G.W. (1990). Characteristics of transport in an acidic and metal-rich mountain stream based on a lithium trace injection and simulations of transient storage. *Water Resources Research*, v. 26, no. 5, pp. 989–1000.

Bertrand, D. (1952). The distribution of lithium in phanerogams. *Weekly reports of the Meetings of the Academy of Sciences*, v. 234, January–June, pp. 2102–2104.

Cade, J.F.J. (1949). Lithium salts in the treatment of psychotic excitement. *Medical Journal of Australia*, v. 2, no. 36, pp. 349–352.

Canon, J.F. (1975). *The Science of Total Environment*. Lancaster, PA: Technomic Publishing Company.

Cannon, H.L., Harms, T.F., and Hamilton, J.C. (1975). Lithium in unconsolidated sediments and plants of the Basin and Range Province, southern California and Nevada. U.S. Geological Survey Professional Paper 918, 23 p.

Domingo, J.L. (1994). Metal-induced developmental toxicity in mammals—A review. *Journal of Toxicology and Environmental Health*, v. 42, no. 2, pp. 123–41.

Drugs.com (2013). Lithium. Accessed 01/05/23 @ http://www.drugs.com/pro/lithium.html.

ESPI Metals (1993). Material safety data sheet for lithium oxide: ESPI meals: Material Safety Data Sheet. CAS number 12057-24-8/. Accessed 01/05/23 @ https://www.espimetals.com/index.php/msds/650-lithium-oxide.

Fischer, E., and Molnar, L. (1997). Growth and reproduction of Eisenia fetida (Oligochaeta, Lumbricidae) in semi-natural soil containing various metal chlorides. *Soil Biology and Biochemistry*, v. 29, nos. 3–4, pp. 667–670.

Garrett, D.E. (2004). *Handbook of lithium and natural calcium chloride* (1st ed.). Boston, MA, Elsevier, 476 p.

Heier, K.S., and Billings, G.K. (1978), Lithium, in Wedepohl, K.H., ed. *Handbook of geochemistry*, v. 2, pt. 1. Berlin, Springer-Verlag, pp. 3-A-1 to 3-A-7.

Kesler, T.L. (1942). *The tin-spodumene belt of the Carolinas, a preliminary report*. Washington, DC, U.S. Geological Survey Bulletin 936-J p 245–269, 5 plates.

Kesler, T.L. (1955). The Kings Mountain area, in Russell, R.J., ed. *Guides to southeastern geology*. New York, Geological Society of America, pp. 374–387, sketch maps.

Kesler, T.L. (1961). Exploration of Kings Mountain pegmatites. *Mining Engineering*, v. 13, no. 9, September, pp. 1063–1068.

Ketter, T.A. (2010). Diagnostic features, prevalence, and impact of bipolar disorder. *The Journal of Clinical Psychiatry*, v. 71, no. 6, p. e14.

Kszos, L.A., and Stewart, A.J. (2003). Review of lithium in the aquatic environment—Distribution in the United States, toxicity and case example of groundwater contamination. *Ecotoxicology*, v. 12, no. 5, pp. 439–447.

Lenntech (2007). *Lithium and water—Reaction mechanisms environmental impact and health effects*. Delft, Lenntech. Accessed 01/05/23 @ http://www.enntech.com/elements-and-water/lithium-and-water.htm.

Leonard, A.J., Hantson, P.E., and Gerber, G.B. (1995). Mutagenicity, carcinogenicity and teratogenicity of lithium compounds. *Mutation Research/Reviews in Genetic Toxicology*, v. 339, no. 3, pp. 131–137.

Lombardi, O.W. (1963). *Observations on the distribution of chemical elements in the terrestrial saline deposits of Saline Valley, California*. China Lake, CA, U.S. Naval Ordinance Test Station, 42 p.

Mitchell, P.B., and Hadzi-Pavlovic, D. (2000). Lithium treatment for bipolar disorder. *Bulletin of the World Health Organization*, v. 78, no. 4, pp. 515–517.

Mitchell, R.L. (1964). Trace elements in soils, chap. 8 of Bear, F.E., ed. *Chemistry of the soil* (2nd ed.). New York, Reinhold, pp. 320–368.

Mouawad, J., and Drew, C. (2013). Praised but fire prone, battery fails test in 787. *The New York Times*, January 17, p. A1. Accessed 01/06/23 @ http://www.nytimes.com/2013/01/18/business/inside-the787-an-unsettling-risk-for-boeing.html.

NEMA (National Electrical Manufacturers Association (2001). Spent consumer lithium batteries and the environment. Accessed 01/06/23 @ http://www.nema.org/Policy/Environmental-Stewardship/Documents/SpentCosumer-Lithium-Batteries-and-the-environment.pdf.

Notter, D.A., Gauch, M., Wimer, R., Wager, Pl, Stamp, A., Sah, R, and Althaus, H.J. (2010). Contribution of Li-ion batteries to the environmental impact of electric vehicles. *Science and Technology*, v. 44, no. 17, pp. 6550–6556.

Schrauzer, G.N. (2002). Lithium—Occurrence, dietary intakes, nutritional essentially. *Journal of the American College of Nutrition*, v. 21, no. 1, pp. 14–21.

Shacklette, H.T. Boerngen, J.G., Cahill, J.P. and Rahill, R.L. (1977). Lithium in surficial materials of the conterminous United States and partial data on cadmium. U.S. Geological Survey Circular 673, 8 p.

Sievers, M.L., and Canon, H.L. (1973). Disease patterns of Pima Indians of the Gila River Indian Reservation of Arizona in relation to the geo-chemical environment. *University of Missouri Symposium on Trace elements in Environmental Health*, v. 7, pp. 57–61.

Sigma-Aldrich (2014). Lithium oxide: Sigma-Aldrich Material Safety Data Sheet, product number 374725, revision date June 28, 2014. Accessed 01/06/23 @ http://www.sigmaaldrichy.com/safety-center.html.

Strobusch, A.D., and Jefferson, J.W. (1980). The checkered history in medicine. *Pharmacy in History*, v. 22, no. 2, pp. 72–76.

U.S. Environmental Protection Agency, Office of Transportation and Air Quality (2011). Greenhouse gas emissions from a typical passenger vehicle: EPA-420-F-11-041, 5 p. Washington, DC, USEPA.

28 Safety Concerns

NOAA SAFETY NOTICE

The following section presents NOAA's April (2020) Lithium Battery Safety Notice. Keep in mind that although this notice is incorporated into the *Small Boat Operations Manual*, it points out the hazards of lithium and the proper safety guidelines in a general and applicable way.

NOAA's Lithium Battery Safety Notice

This safety notice is a supplement to the previously issued September 19, 2019 notice regarding the use and charging of lithium batteries on small boats. There is an increased use of battery-powered devices to support small boat operations and the deployment of higher voltage mission devices is becoming more common. The fleet has had a number of incidents caused by lithium batteries overheating. This warrantees a risk assessment and mitigation plan.

The Small Board Safety Board and Program Office are requiring immediate implementation of the following:

Universal Policy on Lithium Battery Use on NOAA Small Boats

- Distinguish between consumer devices (cell, laptop, cordless tools, and radios) and mission devices (higher voltage science gear, submerged devices, autonomous vehicles, and custom-built instruments).
- Maintain an inventory of standard consumer devices and their locations.
- Identify projects that include mission equipment powered by lithium batteries.
- Incorporate lithium battery fire hazards and response into existing fire drills and training.
- Explain this battery policy and local procedures during vessel orientations.

Requirements for Consumer Devices

- Prohibit unattended charging of consumer devices in berthing spaces.
- Use only chargers that are compatible with the device as defined by the manufacturer.
- Devices must be attended during charging on any surface that is potentially flammable.

DOI: 10.1201/9781003387879-35

- Require regular inspection (monthly, each use, etc.) of battery-powered devices.
- Remove from service, and the vessel, battery cases or devices that are— physically damaged, swollen, cracked, or show evidence of overheating.
- Do not store or charge devices in proximity to flammable materials.

REQUIREMENTS FOR MISSION DEVICES

- Require device-specific procedures and responsibilities for proper handling, charging, and monitoring.
  - Identify any custom/prototype chargers that do not have rate, over-current, of heat monitoring features and consider them to be high risk devices.
- Identify and maintain areas suitable for high-risk recharging and service.
  - These areas require a posting, monitored charging only, and a means of fire containment. Containment may include the use of charging bags, metal dividers, or cans.
  - Location should be dry, open air and allow for overboard jettison when-ever possible.
  - Location must not be near flammable surfaces in in areas that would impede egress.
  - A suitable fire extinguisher shall be located within close proximity of any designated charging station—Class B or ABC rating. Provide post-ing and instruction on fighting a lithium battery fire.
- Do not store or charge devices in proximity to flammable material storage.

Note: In addition, VOCs and program leads shall review operations that require the use of lithium batteries and develop supplemental procedures for any specialized operation not covered under this policy.

Note: Lithium batters are hazardous materials and are subject to DOT's Hazardous Materials Regulations (HMR; 49 CFR Pars 171–180). This includes packaging and standard hazard communication requirements (e.g., markings, labels, shipping papers, and emergency response information) and hazmat employee training require-ments. Hazard communication requirements are found in part 172 of the HMR and requirements specific to lithium batteries are found in 49 CFR section 173.185.

HAZARD COMMUNICATION STANDARD

Hazard Communication?

You mean OSHA's 29 CFR 1910.1200?

Why?

Yes, and Yes, and the why is that it is the law of the land (in the United States). Simply stated, lithium is a hazardous substance that can do considerable damage if mishandled—fire and explosion hazard. Thus, if a substance (including lithium) is harmful to those who might handle, be exposed to, or purchase lithium-based prod-ucts, OSHA's Hazard Communication (HAZCOM) Standard (29 CFR 1910.1200) requires the manufacturer to comply with the standard. The point is if a worker, a passerby, or any other person could potentially be exposed to a hazardous substance they need to know—there must be some type of signage of the warning type that anyone and everyone can see and can be aware of. The bottom line: People have a right to know, otherwise there could be catastrophic results. Consider, for example, the following instance:

Juju

The day rose heavy and hot, but the wind whispered in the field beyond the sod house, as if murmuring delightful secrets to itself. A light breeze entered the open windows and gently touched those asleep inside. A finger of warmth, laden with the rich, sweet odor of the earth, lightly touched Juju's cheek—rousing her this morning, as it had so often in her 9 years of life. On most days, Juju would lay on her straw mat and daydream, languishing in the glory of waking to another day on Mother Earth. But nothing was normal on this morning. This day was different—full of surprises and excitement. Juju, and her mother, Lanruh, were setting out on a new adventure today—and Juju couldn't wait.

As she stood at the foot of her make-shift bed, Juju swiftly tucked the folds of thin fabric around her slender waist, and let the soft cloth fall softly to hang to her feet. She pulled her straight black hair tight in a knot at the back of her neck, before she draped the end of the sari over her head.

While Juju dressed, Lanruh performed the same ritual in her small room, next to Juju's. Lanruh was excited about the day's events, too—She knew Juju was thrilled, and she always delighted in her daughter's pleasure and excitement, when experiencing every new event. Lanruh chuckled to herself, as she remembered the many times over the last few years, that Juju had begged to be included, when Lanruh made her trips to the Grand Market in town. Lanruh understood Juju's excitement. Going into the town, taking in all the sights, and sounds of the market, thrilled Lanruh, too.

As they stepped out of the sod house, and onto the dirt road, the scented breeze that had touched Juju's cheek earlier greeted them. They walked together, hand-in-hand, toward town, 3 km to the south.

Juju bubbled with anticipation, but she held it in, presenting the calm, serene face expected of her. Even so, every nerve in her young body reverberated with excitement.

As they walked along the road, Juju, strived to see all the new landscapes, and was fascinated by everything she saw, and she thrilled to this extension of her, here-to-fore, small world. People and cattle were everywhere —she had never seen so many of either! Her world had grown, suddenly—and it felt good to be alive.

As they neared town, Juju could see the tall buildings. How big and imposing they were—and so many of them! In town, in places they passed, some of the streets were actually paved. Juju had never seen paved streets! This trip to town was her first city experience, and she was enthralled by all the strange and wonderful sights. As they walked along the street leading to the marketplace, Juju was over-awed by the tall buildings and warehouses. "What could they all be used for?" she wondered. Some of them had sign boards above their doors, but little good that did for Juju—she couldn't read.

The steady, gentle, following breeze had escorted Juju and Lanruh, since they left home, and it was still with them as they turned toward the market. Juju's eyes sparkled with excitement as she saw the entrance to the market, and the throngs of bustling people ahead, and she could barely contain her excitement.

Suddenly, with one deep breath of that sweet air, (*was it the same sweet air that had touched her into waking only 2 hours earlier?*) Juju began coughing. She clutched her throat with both hands, her eyes filled with tears of fright, and she fell to her knees in sudden agony. Her mother, fell as well, gasping for air. The breeze that had begun their day of excited anticipation, now ended it—delivering an agent of death. But Juju didn't have time to realize what was happening. She couldn't breathe. She couldn't do anything—except die—and she did.

Juju, Lanruh, and over two thousand other souls in the area, died within a very few minutes, having breathed in that silent, deadly breeze.

Those who died that day (December 3, 1984) never knew what killed them. The several hundred others who died soon after did not know what killed them, either.

The several thousand inhabitants who lived near the marketplace, close by the industrial complex, where the pesticide factory operated, near the chemical spill, near the release point of that deadly toxin—knew little, if any, of this. They knew only death, and the killing sickness which invaded their town on that sorry day.

Those who survived the fatal invader on that day were later told that a deadly chemical had killed their families, their friends, their neighbors, and their acquaintances. They were killed by a chemical spill that today is infamous in the journals of hazardous materials incidents.

Today, this incident is studied by everyone who has anything to do with chemical production and handling operations. We know it as Bhopal.

Those who died knew nothing of the disaster—and their deaths were the result.

F. R. Spellman (1998)

HAZARD COMMUNICATION

The Bhopal Incident, the ensuing chemical spill, and the resulting tragic deaths and injuries are well known. Now however, not all of the repercussions—and the lessons learned—from this incident are not as well known. After Bhopal, there arose a worldwide outcry.

"How could such an incident occur? Why wasn't something done to protect the inhabitants? Weren't there safety measures taken, or in place, to prevent such a disaster from occurring?"

Lots of questions, few answers. The major problem was later discovered to be a failure to communicate. That is, the workers, residents, and visitors had no idea of how dangerous a chemical spill could be. Far too many found out the hard way.

In the United States, these questions, and others, along with "after the fact findings" were bandied around and talked about by the President and by Congress. Because of Bhopal, Congress took the first major step to prevent such incidents from occurring in the United States. What Congress did was to direct OSHA to take a close look at chemical manufacturing in the United States to see if a Bhopal-type incident could occur in this country. OSHA did a study and then reported to Congress that a Bhopal-type incident in the United States was very unlikely.

Tragically, within only a few months of OSHA's report to Congress, a chemical spill occurred, similar to Bhopal, but fortunately, not deadly (*no deaths, but 100+ people became ill*), in the town of Institute, West Virginia.

DID YOU KNOW?

Exposure to hazardous chemicals is one of the most serious dangers facing American workers today, and too many workers may not even understand the risk that they're taking, when working with chemicals.

Needless to say, Congress was upset. Because of Bhopal and the Institute, West Virginia fiascoes, OSHA mandated its Hazard Communication Program, 29 CFR 1910.1200 in 1984. Later, other programs like SARA (Superfund) Title III, reporting requirements for all chemical users, producers, suppliers, and storage entities, were mandated by USEPA.

There is no "*all-inclusive list*" of chemicals covered by the HazCom Standard; however, the regulation refers to "any chemical which is a physical or health hazard." Those specifically deemed hazardous include:

- Chemicals regulated by OSHA in 29 CFR Part 1910, Subpart Z, Toxic and Hazardous Substances
- Chemicals included in the American Conference of Governmental Industrial Hygienists' (ACGIH) latest edition of Threshold Limit Values (TLVs) for *Chemical Substances and Physical Agents in the Work Environment*
- Chemicals found to be suspected, or confirmed carcinogens by the National Toxicology Program, in the *Registry of Toxic Effects of Chemical Substances*, published by NIOSH, or appearing in the latest edition of the *Annual Report on Carcinogens,* or by the International Agency for Research on Cancer, in the latest editions of its IARC *Monographs.*

Congress decided that those personnel working with, or around, hazardous materials, "*had a right to know*" about those hazards. Thus, OSHA's Hazard Communication standard (HCS) was created. The HCS is, without a doubt, the most important regulation in the communication of chemical hazards to employees. Moreover, because OSHA's Hazard Communication is a dynamic (*living*) standard, it has been easily amendable and adjusted to comply with ongoing, worldwide changes in an effort to make employer and worker chemical safety compliance requirements, more pertinent, and applicable.

Considering this on-going desire for currency and applicability, Federal OSHA published a revised HCS, (*HazCom*) on March 26, 2012, which aligns with the *United Nation's Globally Harmonized System of Classification and Labeling of Chemicals.*

This revision affects how chemical hazards are classified, the elements incorporated into a label, and the format of the safety data sheet (SDS). In addition, some terminology and several definitions have changed, including the definition of a hazardous chemical.

Under its HCS (*more commonly known as "HazCom" or the "Right to Know Law"*), OSHA requires employers who use, or produce chemicals on the worksite, to inform all employees of the hazards that might be involved with those chemicals. HazCom says that employees have the right to know what chemicals they are handling or may be exposed to. HazCom's intent is to make the workplace safer. Under the HazCom Standard, the employer is required to fully evaluate all chemicals on the worksite for possible physical and health hazards. All information relating to these hazards must be made available to the employee 24 hours each day. The standard is written in a performance manner, meaning that the specifics are left to the **employer** to develop.

HazCom also requires the *employer* to ensure proper labeling of each chemical, including chemicals that might be produced by a *process* (process hazards). For example, in the wastewater industry, deadly methane gas is generated in the waste stream. Another common wastewater hazard is the generation of hydrogen sulfide, (*which produces the characteristic rotten-egg odor*) during the degradation of organic substances in the wastestream and can kill quickly. OSHA's HazCom requires the employer to label methane, and hydrogen sulfide hazards, so that workers are warned, and safety precautions are followed.

Labels must be designed to be clearly understood by all workers. Employers are required to provide both training and written materials, to make workers aware of what they are working with, and what hazards they might be exposed to. Employers are also required to make SDS available to all employees. An SDS is a fact sheet for a chemical which poses a physical or health hazard in the work place. An SDS must be in English and contain the following information:

- Identity of the chemical (label name)
- Physical hazards
- Control measures
- Health hazards
- Whether it is a carcinogen
- Emergency and first aid procedures
- Date of preparation of the latest revision
- Name, address, and telephone number of manufacturer, importer, or other responsible parties

Blank spaces are not permitted on an SDS. If relevant information in any one of the categories is unavailable at the time of preparation, the SDS must indicate that no information was available. Your facility must have an SDS for each hazardous chemical it uses. Copies must be made available to other companies working on your worksite (*outside contractors, for example*), and they must do the same for you. The facility hazard communication program must be in writing and, along with an SDS, be made available to all workers, 24 hours each day/each shift. Information contained in a typical SDS for lithium (generic by author) is presented in the following.

Lithium Safety Data Sheet (Example)

1. Product and Company Identification

Product name:	Lithium
Chemical Formula:	Li
CAS# (Chemical Abstracts Service)	7439-93-2
Contact Information	manufacturer's contact information
Emergency assistance	Telephone number listed here

2. Hazards Information

Emergency Overview: Globally Harmonized System (GHS) Classification in accordance with 29 CFR 1910 (OSHA HCS). Substances and mixtures, which in contact with water, emit flammable gases (Category 1—highest level of risk), H260

Skin Corrosion (category 1A), H314 (a skin sensitizer)

Serious eye damage (Category 1), H318—causes irreversible effects on the eye

HMIS Rating

Health hazard	3
Flammability	2
Physical Hazard	2

NFPA Rating

Health Hazard	3
Fire Hazard	2
Reactivity Hazard	2
Special Hazard	W (water and lithium do not mix—avoid water contact

GHS Label Elements, Including Precautionary Statements

Pictogram and Signal Danger

Hazard Statement(s)

H260	In contact with water releases flammable gases which may ignite spontaneously.
H314	Causes severe burns and eye damage.

Precautionary Statement(s)

P210	Keep away from Heat/sparks/open flames/surfaces. No Smoking.
P231 + P232	Handle under inert gas. Protect from moisture.
P260	Do not breathe dust/fume/gas/mist/vapors/spray
P264	Wash skin thoroughly after handling
P280	Wear protective gloves/protective clothing/eye protection/face protection.
P301 + P330 + P331	IF SWALLOWED: Rinse mouth. Do NOT induce Vomiting.
P330 + P363 + P353	IF ON SKIN (or hair): Take off immediately all contaminated clothing. Rinse skin with water/shower.
P304 + P340	IF INHALED: Remove victim to fresh air and keep at rest in a position comfortable for breathing.
P305 + P351 + P338	IF IN EYES: Rinse cautiously with water for several minutes. Remove contact lenses, if present and easy to do. Continue rinsing.
P310	Immediately call a POISON CENTER or doctor/physician.
P321	Specific treatment.
P335 + P334	Brush off loose particles from skin. Immerse in cool water/wrap in west bandages.
P363	Wash contaminated clothing before reuse.
P370 + P378	In case of fire: Use dry sand, dry chemical or alcohol-resistant foam to extinguish.
P402 + P404	Store in a dry place. Store in closed container.
P405	Store locked up.
P501	Dispose of contents/container to an approved waste disposal plant.
Bottom line:	Reacts violently with water.

(Continued)

Lithium Safety Data Sheet (Example)

3. Composition/Information on Ingredients

Substance Name   **Lithium**
Formula   Li
Molecular Weight   6.94 g/mol
CAS-No.   7439-93-2
Component   Li 99.9%
Hazardous components   Water-reactive; skin corrosive; eye damage

4. First Aid Measures

Description of First Aid Measures

General advice—Consult a physician. Show this SDS to physician. Move out of dangerous area

If inhaled—If breathed in, move person into fresh air. If not breathing, give artificial respiration. Consult physician.

In case of skin contact—Take off contaminated clothing and shoes immediately. Wash off with soap and plenty of water. Consult a physician.

In case of eye contact—Check for and remove any contact lenses. Rinse thoroughly with plenty of water for at least 15 minutes and consult a physician. Continue rinsing eyes during transport to hospital.

If Swallowed—Do NOT induce vomiting. Never give anything by mouth to an unconscious person, Rinse mouth with water. Consult a physician.

5. Firefighting Measures

Conditions of flammability—May burn in presence of air or emit a flammable gas in the presence of water or water vapor. Keep away from heat/sparks/open flame/hot surface/air/water. No Smoking.

Extinguishing media—Use approved class D extinguishers of smother with dry sand, dry ground limestone, or dry clay. DO NOT use water, foam, or carbon dioxide.

Special hazards arising from the substance mixture—Lithium oxides.

Advice for fire fighters—Wear self-contained breathing apparatus for firefighting if necessary.

Hazardous combustion products—Hazardous decomposition products formed under fire conditions—lithium oxides.

6. Accidental Release Measures

Do not let produce enter drains. Personal precautions, protective equipment and emergency procedures—Use personal protective equipment (PPE). Avoid breathing vapors, mist or gas. Ensure adequate ventilation. Remove all sources of ignition. Evacuate personnel to safe areas.

Environmental precautions—Prevent further leakage or spillage if safe to do so. Do not let product enter drains.

Methods and materials for containment and cleaning up—sweep up with shovel. Contain spillage, and then collect with an electrically protected vacuum cleaner or by wet brushing and place in container for disposal according to local regulations. Do not flush with water. Keep in suitable closed containers for disposal.

7. Handling and Storage

Precautions for safe handling—Avoid contact with skin and eyes. Avoid inhalation of vapor or mist. Avoid formation of dust and aerosols. Provide appropriate exhaust ventilation at places where dust is formed. Keep away from sources of ignition—No Smoking. Take measures to prevent the buildup of electrostatic charge.

Conditions for safe storage, including any incompatibilities—Store under argon. Handle under argon. Keep container tightly closed in a dry and well-ventilated place. Never allow product to get in contact with water during storage. Do not store in glass. Air and moisture sensitive.

(Continued)

Lithium Safety Data Sheet (Example)

8. Exposure Control/Personal Protection

Control parameters

Components with workplace control parameters—contains no substances with occupational exposure limit values.

Exposure controls

Appropriate engineering controls—handle in accordance with good industrial hygiene and safety practice. Wash hands before breaks and at the end of workday.

PPE

Eye/face protection—face shield and safety glasses. Use equipment for eye protection tested and approved under appropriate government standards such as NIOSH (US) or EN 166(EU).

Skin protection—handle with gloves. Gloves must be inspected before use. Use proper glove removal techniques (without touching glove's out surface) to avoid skin contact with this product. Dispose of contaminated gloves after use in accordance with applicable laws and good laboratory practices. Wash and dry hands.

Body protection—complete suit protecting against chemicals, flame retardant antistatic protective clothing. The type of protective equipment must be selected according to the concentration and amount of the dangerous substance at the specific workplace.

Respiratory protection—where risk assessment shows air-purifying respirators are appropriate use a full-face respirator with multipurpose combination (US) or type ABEK (EN 14387) respirator cartridges as backup to engineering controls. If the respirator is the sole means of protection, use a full-face supplied air respirator. Use respirators and components tested and approved under appropriate government standards such as NIOSH (US) or CEN (EU).

Control of environmental exposure—prevent further leakage or spillage if safe to do so. Do not let product enter drains.

9. Physical and Chemical Properties (currently available information)

Selected Basic Physical and Chemical Properties (as known at present)

Appearance   Solid

Melting point/freezing point   180°C (356°F)

Initial boiling point and boiling range   1,342°C (2,448°F)

Vapor pressure   1 hPa (1 mmHg) at 723°C (1,333°F)

Relative density   0.534 g/mL at 25°C (77°F)

10. Stability and Reactivity

Chemical stability—stable under recommended storage conditions.

Possibility of hazardous reactions—reacts violently with water.

Conditions to avoid—exposure to moisture.

Incompatible materials—forms shock-sensitive mixtures with certain other materials, iron and iron salts, heavy metals, phosphorus, sulfur compounds, oxygen, nickel. Do not store near acids, metals, chlorinated solvents, water, nitrogen.

11. Toxicological Information

Large doses of lithium ion have caused dizziness and prostration and can cause kidney damage if sodium intake is limited. Dehydration, weight loss, dermatological effects, and thyroid disturbances have been reported. Central nervous system effects that include slurred speech, blurred vision, sensory loss, ataxia, and convulsions may occur. Diarrhea, vomiting, and neuromuscular effects such as tremor, clonus, and hyperactive reflexes may occur as result of repeated exposure to lithium.

12. Ecological Information

No data available

(Continued)

Lithium Safety Data Sheet (Example)

13. Disposal Considerations

Waste Treatment Methods

Product—burn in a chemical incinerator equipped with an afterburner and scrubber but exert extra care in igniting as this material is highly flammable. Dissolve or mix the material with a combustible solvent and burn in a chemical incinerator equipped with an afterburner and scrubber. Contaminated packaging should be disposed of as unused product.

14. Transport Information

DOT (US)

UN Number: 1415; Class 4.3; Packing group: 1

IMDG and IATA are basically the same

15. Regulatory Information

SARA 302 Components—material is not subject to the reporting requirements of SARA Title III, Section 302.

SARA 313 Components—material does not contain any chemical components with CAS numbers that exceed the threshold reporting levels established by SARA Tile III, Section 313.

SARA 311/312 Hazards—fire hazard, acute health hazard.

16. Other Information

Eye damage   Serious eye damage

H260   In contact with water releases flammable gases which may ignite spontaneously.

H314   Causes severe skin burns and eye damage.

H318   Causes severe eye damage.

Skin Corrosion   Skin corrosion

Water-reactivity   Substances and mixtures, which in contact with water, emit flammable gases.

COMMUNICATION FOR WORKER SAFETY AND HEALTH

To provide better worker protection from hazardous chemicals, and to help American businesses compete in a global economy, OSHA has revised its Hazardous Communication (**HazCom**) standard to align with the **United Nations' Globally Harmonized System of Classification and Labeling of Chemicals**—referred to as *GHS*. This *GHS* incorporates the quality, consistency, and clarity of hazard information that workers receive, by providing harmonized criteria for classifying, and labeling, all hazardous chemicals, and for preparing SDSs for these chemicals.

The *GHS* system is an innovative approach that has been developed through international negotiations and embodies the knowledge gained in the field of chemical hazard communication since the HazCom standard was first introduced in 1983. Simply, *HazCom*, along with *GHS,* means better communication to reduce chemical hazards for workers on the job.

Benefits of HAZCOM with GHS1

Practicing Occupational Safety and Health professionals are familiar with OSHA's original 1983 HCS. Many are now becoming familiar with the phase-in of the new combined *HazCom* and **GHS** standard. The first thing they learn is that the **GHS** is an international approach to hazard communication, providing agreed upon criteria for classification of chemical hazards, and a standardized approach to labeling elements, and SDSs. The *GHS* was negotiated in a multi-year process by hazard communication experts from many different countries, international organizations, and stakeholder groups. It is based on major existing systems around the world, including OSHA's HCS and the chemical classification, and labeling systems, of other US agencies.

The result of this negotiation process is the United Nations' document entitled "Globally Harmonized System of a Classification and Labeling of Chemicals, "commonly referred to as **The Purple Book**. This document provides harmonized classification criteria for health, physical, and environmental hazards of chemicals. It also includes standardized label elements that are assigned to these hazard classes, and categories, and provides the appropriate signal words, pictograms, and the hazard, and precautionary statements necessary to convey the types of hazards to users. A standardized order of information for SDSs is also provided. These recommendations can be used by regulatory authorities, such as OSHA, to establish mandatory requirements of hazard communication, but do not constitute a model regulation.

OSHA's motive to modify the *HCS* was to improve the safety and health of workers, through more effective communications on chemical hazards. Since it was first promulgated in 1983, the *HCS* has provided employers, and employees, extensive information about the chemicals in their workplaces. The original standard is performance-oriented, allowing chemical manufacturers, and importers, to convey information on labels, and on material data sheets, in whatever format they choose. While the available information has been helpful in improving employee safety and health, a more standardized approach to classifying the hazards, and for conveying the information, will be more effective, and provide further improvements in American workplaces. The **GHS** provides such a standardized approach, including detailed criteria for determination of what hazardous effects a chemical can cause, as well as standardized label elements assigned by hazard class, and category. This will enhance both employer and worker comprehension of the hazards, which will help to ensure appropriate handling, and safe use, of workplace chemicals. In addition, the SDS requirements establish an order of information that is standardized. The harmonized format of the SDSs will enable employers, workers, health professionals, and emergency responders, to access the information more efficiently, and effectively, thus increasing their utility.

Adoption of the *GHS* in the US, and around the world, will also help to improve information received from other countries--since the US is both a major importer, and exporter of chemicals. American workers often see labels and SDSs form other countries. The diverse, and sometimes conflicting, national, and international requirements can create confusion among those who seek to use hazard information more effectively. For example, labels and SDSs may include symbols, and hazard statements, that are unfamiliar to readers, or may not be well understood. Containers may be labeled with such a large volume of information, (*overkill*) that important statements are not easily recognized. Given the differences in hazard classification criteria, labels may also be incorrect when used in other countries. If countries around the world adopt the *GHS*, these problems will be minimized, and chemicals crossing borders will have consistent information, thus improving communication globally.

Phase-In Period for the HAZARD COMMUNICATION STANDARD

During the phase-in period (2013–2016), employers were required to be following either the existing HCS, or the revised HCS, or both. OSHA recognized that hazard communication programs would go through a period where both labels, and SDS sheets, would be present in the workplace, under both standards. This was considered acceptable, and employers were not required to maintain two sets of labels, and SDS sheet, for compliance purposes.

It is important to point out that prior to OHSA's effective compliance date for full implementation of the revised HCS, that employee training was required to be conducted. This was the case, because American workplaces received the new SDS, and labeling requirements before the full compliance date was met. Thus, employees needed to be trained early to enable them to recognize, and understand, the new label elements (*i.e., pictograms, hazard statements, precautionary statements, and signal words*) and the SDS format.

Major Changes to the hazard communication standard

There are three major areas of change in the modified HCS: in hazard classification, labels, and SDSs.

- **Hazard classification**—The definitions of hazard have been changed to provide specific criteria for classification of health and physical hazards, as well as classification of mixtures. A result of these specific criteria is to ensure that evaluations of hazardous effects are consistent across manufacturers, and that labels and SDSs are more accurate.
- **Labels**—Chemical manufacturers and importers will be required to provide a label that includes a harmonized signal word, pictogram, and hazard statement for each hazard class, and category. Precautionary statements must be provided.
- **SDS**—Will now have a 16-section format.

Note: The **GHS** does not include harmonized training provisions but recognizes that training is essential to achieve an effective hazard communication approach. The revised **HCS**) requires that workers be re-trained within 2 years of publication of the final result, and this would serve to facilitate more effective recognition, and understanding, of the new labels and SDSs.

Hazard Classification

Not all HCS provisions are changed in the revised **HCS**. The revised **HCS** is simply a modification to the existing standard and has been designed to make it universal and worker-friendly. The parts of the standard that did not relate to the **GHS** (*such as the basic framework, scope, and exemptions*) remained largely unchanged. There have been some modifications in terminology, as an effort to more closely align the revised **HCS**, with the language used in the **GHS**. For example, the term "*hazard* *determination*" has been changed to "*hazard classification*" and "*material safety data sheet*" was changed to "*safety data sheet.*"

Under both the current, **HCS** and the **revised HCS**, an evaluation of chemical hazards must be performed considering the available scientific evidence concerning such hazards. Under the current **HCS**, the hazard determination provisions have definitions of *hazard*, and the evaluator had to determine whether, the data on a chemical met those definitions. It is a performance-oriented approach that provides parameters for the evaluation, but not specific, detailed criteria.

The hazard classification approach in the revised **HCS** is quite different. The revised **HCS** has specific criteria for each health and physical hazard, along with detailed instructions for hazard evaluation, and determinations as to whether mixtures or substances are covered. It also establishes both hazard *classes*, and hazard *categories*--for most of the possible effects caused. The classes are divided into categories that reflect the relative severity of the effect. The current **HCS** does not include categories for most of the health hazards covered, so this innovative approach provides additional information that can be helpful in providing the appropriate response to deal with the hazard more effectively. **OSHA** has included the general provisions for hazard classification in paragraph (d) of the revised rule and added extensive appendices that address the criteria for each health, or physical effect.

LABEL CHANGES UNDER THE REVISED HCS

Under the current **HCS**, the label preparer must provide the identity of the chemical, and the appropriate hazard warnings. This may be done in a variety of ways, and the method to convey the information is left to the preparer. Under the *revised HCS*, once the hazard classification is completed, the standard specifies what information is to be provided for each hazard class, and category. Labels will require the following elements:

- **Pictogram**: A symbol plus other graphic elements, such as a border, background pattern, or color that is intended to convey specific information about the hazards of a chemical. Each pictogram consists of a specified symbol on a white background, within a red square set on a point, or diamond shape (*i.e., a white background with a border*) (see Figure 28.1). There are nine pictograms under the **GHS**. However, only eight pictograms are required under the **HCS**. Note that the Environment pictogram, shown in Figure 26.1, is not mandatory; however, the other eight are mandatory.
- **Signal words**: a signal word on the label, is used to indicate the relative level of severity of a hazard and alerts the reader to a potential hazard. The signal words used are "danger" and "warning" (see Figure 28.2). **"Danger"** is used for the more severe hazards, while **"Warning"** is used for less severe hazards.
- **Hazard statement**: a statement assigned to a hazard class and category, that describes the nature of the hazard(s) of a chemical including, where appropriate, the degree of hazard.
- **Precautionary Statement**: a phrase that describes recommended measures to be taken to minimize, or prevent, adverse effects resulting from exposure to a hazardous chemical or   due to the improper storage, or handling of a hazardous chemical.

HAZCOM STANDARD PICTOGRAMS

Health Hazard	Flame	Exclamation Mark
• Carcinogen • Mutagenicity • Reproductive Toxicity • Respiratory Sensitizer • Target Organ Toxicity • Aspiration Toxicity	• Flammables • Pyrophorics • Self-Heating • Emits Flammable Gas • Self-Reactives • Organic Peroxides	• Irritant (skin and eye) • Skin Sensitizer • Acute Toxicity (harmful) • Narcotic Effects • Respiratory Tract Irritant • Hazardous to Ozone Layer (Non-Mandatory)
Gas Cylinder	**Corrosion**	**Exploding Bomb**
• Gases Under Pressure	• Skin Corrosion/ Burns • Eye Damage • Corrosive to Metals	• Explosives • Self-Reactives • Organic Peroxides
Flame Over Circle	**Environment** (Non-Mandatory)	**Skull and Crossbones**
• Oxidizers	• Aquatic Toxicity	• Acute Toxicity (fatal or toxic)

FIGURE 28.1   Global Harmonized Labels

In the revised HCS, OSHA is lifting the stay on enforcement regarding the provision to update labels when additional information on hazards becomes available. Chemical manufacturers, importers, distributors, or employers who become newly aware of any significant information regarding the hazards of a chemical, shall revise the labels for the chemical within 6 months of becoming aware of the latest information. If the chemical is not currently produced or imported, the chemical manufacturer, importer, distributor, or employer shall add such information to the label before the chemical is shipped or introduced into the workplace again.

The current standard provides employers with flexibility regarding the type of system to be used in their workplaces, and OSHA has retained that flexibility in the revised **HCS**. Employers may choose to label workplace containers either with the same label that would be on the shipped container for the chemical under the revised rule, or with label alternatives that meet the requirements for the accepted standard. Alternative labeling systems such as, the National Fire Protection Association (NFPA) 704 Hazard Rating, and the Hazardous Material Identification System (HMIS), are permitted for workplace containers. However, the information supplied on these labels must be consistent with the revised HCS, (i.e., no conflicting hazard warnings or pictograms).

SDS Changes under the Revised HCS

The information required on the (Material) **SDS** will remain essentially the same as in the current standard (**HazCom 1994**). HazCom 1994 indicates what information must be included on an SDS, but does not specify a format for presentation, or order of information. The revised HCS (2012) requires that the information on the SDS be presented in a specified sequence. The revised SDS should contain 16 headings (Table 28.1).

FIGURE 28.2   Sample signal word labels

TABLE 28.1
Minimum Information for an SDS

1. Identification of the substance, or mixture, and of the supplier	• GHS product identifier • Other means of identification • Recommended use of the chemical and restrictions on use • Supplier's details (including name, address, phone number, etc.) • Emergency phone number
2. Hazards Identification	• GHS classification of the substance/mixture, and any national, or regional information • GHS label elements, including precautionary statements. (*Hazard symbols may be provided as a graphical reproduction of the symbols in black and white or the name of the symbol, e.g., flame, skull, and crossbones*) • Other hazards which do not result in classification (*e.g., dust explosion hazard*) or are not covered by GHS
3. Composition/Information on Ingredients	**Substance** • Chemical identity • Common name, synonyms, etc. • CAS number, EC number, etc. • Impurities and stabilizing additives which are themselves, classified, and which contribute to the classification of the substance **Mixture** • The chemical identity and concentration, or concentration ranges, of all ingredients which are hazardous within the meaning of the GHS and are present above their cutoff levels
4. First aid measures	• Description of necessary measures, included according to the different routes of exposure i.e., inhalation, skin and eye contact, and ingestion • Most important symptoms/effects, acute and delayed • Indication of immediate medical attention and special treatment needed, if necessary
5. Firefighting measures	• Suitable (*and unsuitable*) extinguishing methods • Specific hazards arising from the chemical (*e.g., nature of any hazardous combustion products*) • Special protective equipment and precautions for firefighters
6. Accidental release measures	• Personal precautions, protective equipment and emergency procedures • Environmental precautions • Methods and materials for containment and cleaning up
7. Handling and storage	• Precautions for safe handling • Conditions for safe storage, including any incompatibilities
8. Exposure controls/ personal protection	• Control parameters, e.g., occupational exposure limit values, or biological limit values • Appropriate engineering controls • Individual protection measures, such as PPE
9. Physical and chemical properties	• Appearance (*physical state, color, etc.*) • Odor • Odor threshold • pH

(Continued)

TABLE 28.1 (CONTINUED)
Minimum Information for an SDS

	• melting point/freezing point
	• initial boiling point and boiling range
	• flash point
	• evaporation rate
	• flammability (*solid, gas*)
	• upper/lower flammability, or explosive limits
	• vapor pressure
	• vapor density
	• relative density
	• solubility (*ies*)
	• partition coefficient: n-octanol/water
	• autoignition temperature
	• decomposition temperature
10. Stability and reactivity	• Chemical stability
	• Possibility of hazardous reactions
	• Conditions to avoid (*e.g., static discharge, shock, or vibration*)
	• Incompatible materials
	• Hazardous composition products
11. Toxicological information	Concise, but complete, and comprehensible description of the various toxicological (*health*) effects and available data used to identify those effects, including:
	• Information on the likely routes of exposure (*inhalation, ingestion, skin and eye contact*)
	• Symptoms related to the physical, chemical and toxicological characteristics
	• Delayed and immediate effects and also chronic effects from short- and long- exposure
12. Ecological information	• Ecotoxicity (aquatic and terrestrial, where available)
	• Persistence and degradability
	• Bio accumulative potential
	• Mobility in soil
	• Other adverse effects
13. Disposal considerations	• Description of waste residues and information on their safe handling, and methods of disposal, including the disposal of any contaminated packaging
14. Transportation information	• UN number
	• Transport Hazard class(es)
	• Packing group, if applicable
	• Marine pollutant (*Yes/No*)
	• Special precautions which a user needs to be aware of, or needs to comply with, in connection with transport, or conveyance either within, or outside that premises
15. Regulatory information	• Safety, health and environmental regulations specific for the product in question
16. Other information including information on preparation and revision of SDS	

DID YOU KNOW?

Under the revised HCS, pictograms must have red borders. OSHA believes that the use of the red frame around the white diamond shape, will increase recognition and comprehensibility. Therefore, the red frame is required regardless of whether the shipment is domestic or international. Moreover, the revised HCS requires that all red borders printed on the label have a symbol printed inside it. If OSHA were to allow blank red borders, workers may be confused about what they mean, and become concerned that some information is missing. OSHA has determined that prohibiting the use of blank red borders on labels is necessary to provide the maximum recognition, and impact of warning labels, and to ensure that users do not get desensitized to the warnings placed on labels.

HAZCOM AND LITHIUM: THE BOTTOM LINE

Lithium batteries are generally safe and unlikely to fail, but only so long as there are no defects and the batteries are not damaged. When lithium batteries fail to operate safely or are damaged, they present a fire and/or explosion hazard. Damage from improper use, storage, or charging may also cause lithium batteries to fail.

The real bottom line: Lithium is the new oil, or maybe new gold but with lithium remember safety first—always.

NOTE

1   Based on information from OSHA's (2014).

REFERENCES

OSHA's Hazard Communication Standard. Rockville, MD, Government Institutes, 1998.
OSHA's (2014). *Modification of the Hazardous Communication Standard (HCS) to conform with the United Nations' (UN) Globally Harmonized System of classification and labeling of chemicals (GHS).* Accessed 01/16/15 @ https://www.osha.gov/dsg/hazcom/hazcom-fag.html.
Spellman, F. R., *Surviving an OSHA Audit.* Lancaster, PA, Technomic Publishing Company, 1998.

Epilogue

The purpose of *The Science of Lithium: The New Oil* is to provide the 411 on lithium and lithium-ion batteries and their use and potential future use, which is presently a hot item, a hot topic, and is gaining in world-wide attention and is being seen as a reliable power source "of the future." With the lithium-ion battery, such terms as cathode, anode, and electrolyte are becoming more discussed and recognized.

Let's get back to *"Of the future"*—it is the bottom line of lithium usage for powering electric vehicles and other transportation power sources of the future. What this means is that although lithium use for powering our vehicles is a way to enhance the condition of our environment, there are a few characteristics of lithium usage that have been plagued by safety issues since its introduction in the 1990s. And safety is a concern—when not properly handled, lithium can cause fire and/or explosion. The good news is that the lithium-ion applications where safety is a concern are being addressed and corrected/mitigated to the extent possible.

The jury is still out on whether movement toward using lithium-ion power as the absolute trend for powering our vehicles is destined to replace usage of hydrocarbon fuels—for the purpose of saving our environment. Is it a certainty or not? On the other hand, will fuel-cell power systems be more beneficial?

Are we putting the carriage before the horse? Are we shutting down production of hydrocarbon fuel production before we have a reliable and accessible replacement?

Can we accept a driving range of 350 miles (approximately) provided by lithium-ion batteries or should we wait for technology to develop a fuel cell that can be changed out at change-out locations (gas stations) with a vehicle hypothetical range of 10,000 miles before changeout? Sounds like a dream, doesn't it—but anything is possible. And is changeout of a discharged vehicle fuel cell for another fully "charged" fuel cell is a procedure or process that is no more burdensome than filling up any vehicle-on-wheels gas or diesel fuel tank?

Is the lithium-ion battery the new oil or is it the yet to be developed fuel cell that will really grease the skids—so to speak?

You be the judge.

Index

Note: **Bold** page numbers refer to tables; *italic* page numbers refer to figures.

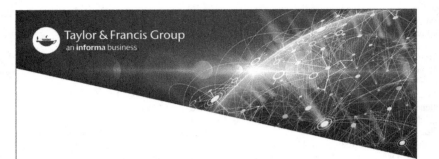

Printed in the United States
by Baker & Taylor Publisher Services